D0014434

FOOD FRAY

Inside the Controversy over
Genetically Modified Food

Lisa H. Weasel

AMACOM

American Management Association
New York • Atlanta • Brussels • Chicago • Mexico City • San Francisco
Shanghai • Tokyo • Toronto • Washington, D.C.

Special discounts on bulk quantities of AMACOM books are available to corporations, professional associations, and other organizations. For details, contact Special Sales Department, AMACOM, a division of American Management Association, 1601 Broadway, New York, NY 10019.
Tel.: 212–903–8316. Fax: 212–903–8083.
Web site: www.amacombooks.org

This publication is designed to provide accurate and authoritative information in regard to the subject matter covered. It is sold with the understanding that the publisher is not engaged in rendering legal, accounting, or other professional service. If legal advice or other expert assistance is required, the services of a competent professional person should be sought.

Library of Congress Cataloging-in-Publication Data

Weasel, Lisa H.
    Food fray : inside the controversy over genetically modified food / Lisa H. Weasel.
        p.   cm.
    Includes bibliographical references and index.
    ISBN-13: 978–0–8144–0164–4
    ISBN-10: 0–8144–0164–3
    1. Genetically modified foods—Political aspects.   I. Title.

TP248.65.F66W43   2009
363.19′2—dc22

                                                    2008028226

Printing number

10   9   8   7   6   5   4   3   2   1

To my daughter Kiran,
with gratitude that we found
each other on this journey

Contents

Preface

"Americans just don't care about genetically modified foods. . . ." This phrase haunted the writing of this book, starting with the initial research phase. And to a limited extent, such a statement may hold a grain of truth. For a segment of the U.S. population—weaned on fast food and addicted to industrialized, highly processed ingredients—questions of how their food is made and where it comes from may be overshadowed by more immediate and pressing food-related outcomes such as obesity, diabetes, and clogged arteries. As one European consumer quoted in Chapter 3 of this book put it, "If Americans are willing to eat McDonald's special sauce, it's no wonder they don't care if it's genetically modified or not."

But any good writer does not take such a dismissal at face value, and both the research for this book as well as recent trends and statistics suggest that it is not so much that Americans don't *care* about genetically modified (GM) food but that they don't *know* about GM food. The most recent public survey conducted by the Pew Charitable Trusts, a respected tracker of biotechnology trends, revealed that 60 percent of Americans believe that they have never eaten GM foods, despite the omnipresence of such foods in a majority of supermarket products for nearly a decade. Moreover, only 38 percent of respondents said they would be likely to eat GM food given the choice, while 54 percent said that they would be unlikely to do so. In obtaining information about GM foods, consumers most often look to their family and friends, trusting government, media, food manufacturers, and the bio-

technology industry the least. At best, a smokescreen of ignorance and uncertainty cloaks the topic of GM food for most Americans.

At the same time, for a growing segment of the U.S. population, issues
of where their food comes from, who controls the food supply, and what
choices they make at the supermarket loom larger than ever. The burgeoning
explosion of the organic food sector in the United States, which expressly
prohibits the presence of more than 5 percent GM content, is testament to
the types of food choices Americans are making that steer away from the GM
trend. And the recent and repeated mass uprisings from consumers against
removing labels from recombinant growth hormone–treated milk in a number of states—discussed in detail in Chapter 8 of this book—suggests that
for at least some products, consumers are willing to act against the use of
recombinant DNA technology in their food, once they are informed about it.

Ironically, one of the ethical arguments that is often used to sway
American consumers in favor of GM foods is not their own self-interest with
regard to these products but the premise that GM foods are needed to help
solve the world hunger problem and feed the world's ever expanding popu-
lous poor. It was this question—a matter of the global *ethics* of GM food,
rather than the global politics—that initially sparked the research for this
book. Would GM food, could GM food, be the answer to the persistent
scourge of famine that strikes with such severity across broad swathes of the
world? And if so, why were opinions so divided and acceptance so controversial in the very places such engineered food might be needed most?

How had Americans and Europeans, despite their common privilege
and relative wealth, come to such different conclusions on the suitability of
these so-called miracle seeds? And how had the debate created such strange
bedfellows? Argentina, for example, joined the United States and Canada in
a World Trade Organization (WTO) lawsuit against Europe over the rejection
of GM foods there. Suddenly, costume-clad, hippy-hugging protesters were
seeing eye-to-eye with a retired director of the American Cancer Society, the
beer company Anheuser-Busch, and the Grocery Manufacturers of America

trade group on the issue of genetically engineering pharmaceuticals into food crops.

To the uninitiated observer, these questions all might seem on the surface to be ones of primarily ethical importance. Are GM foods really a dangerous technology in disguise? Or are they simply misunderstood and thus mistrusted new miracles wasting away on the shelf while regulators twiddle their thumbs and activists indulge the public's fears in vain? Is the rejection of this technology wherever it occurs—whether in Europe, Africa, or Asia—really immoral, as members of the administration of President George W. Bush repeatedly insisted beginning in 2003? Or are such refusals justified, cases of small, brave nations and altruistic activists standing up in the face of questionable foodstuffs being foisted upon them by multinational corporations backed by bullying big-brother superpowers?

Thus, initially, this book was motivated by a desire to examine and interrogate the ethical dimensions of the global arguments surrounding GM food, to bring to the table the voices and perspectives of those so often referred to yet never legitimately heard from, to journey to the heart of the debate and conjure forth a wide chorus of voices grappling with the ethics of GM food. But such a trajectory was not to be, at least not exclusively. For the more I delved into ethics with my respondents—the more I coaxed and encouraged them to talk about the morality and justice surrounding GM foods—the more the untidy and inconvenient topic of politics begged to be let in. Stories of superpowers, whether in the form of wartime bully economies, greedy multinational corporations, or superficial media messages drowning out the more detailed elements of the issue, threaded their way into the interviews and research with persistence.

Perhaps it is a particularly American predicament to presume that ethics and politics have no common denominator, but the interviewees and informants for this research did not share such a view. As I steered our conversations back to the topic of ethics and equity, they insisted on telling stories of government oversight, global power struggles, and the use of food as a political pawn. Media coverage of agricultural biotechnology during this time did not help: Were stories of corporate bribes offered to government regulatory agencies, or of undercover taxpayer subsidies of corporate biotech fines (both of which are detailed between the pages of this book), matters of ethics or politics?

On the surface, this book can certainly be read as a snapshot in time of how different cultures and countries relate to an emergent new manner of producing and growing food, one that though invisible to the naked eye charts a significantly new course in how seeds are created and cultivated. Lying beneath the surface of the story of GM food, though, is a deeper and more persistent narrative of how science, ethics, and politics intersect, not just in the United States and Europe but in the legions of developing nations where some of the strongest debates surrounding GM food are currently taking place.

The story of the global politics of GM food remains an unfinished one, for as much as agricultural biotechnology is indeed here to stay, resistance to its use and application continues to mount. As GM technologies veer off in new directions—such as testing out the use of plants as factories for human and animal drugs and fine-tuning former food crops to make them more amenable as fuel—global reactions will continue to ebb and flow, likely popping up in surprising and unanticipated ways and shaping the course of consumer reaction to these products. The stories presented here represent a starting point, a lens through which some of the more daunting ethical and political consequences of this technology can begin to be grappled with. Hopefully, they will provide ample "food for thought" on the GM debate, regardless of where one's food politics may lie.

Acknowledgments

Books such as this one do not come easily. Taking on a controversial topic with a fast-moving trajectory and an ever-shifting target required the skill, support, and flexibility of a wide range of exemplary individuals.

My university department offered the resilience to ebb and flow with my hectic and often unpredictable travel schedule, as I set off on short notice to investigate the latest GM controversies taking place around the world. Stan Hillman, my department chair at the time, deserves particular acknowledgment for encouraging me, then a junior scholar, to pursue the research that empassioned me, and for arranging my teaching schedule to accommodate my need for travel and writing.

Once on the ground in my various destinations, and even before my arrival, a number of gracious individuals assisted me in seeking out interviews and arranging my itineraries such that I could dig deep into the heart of my subject matter. In Zambia, Beatrice Mwape served as a gracious host and helpful informant. In India, Salome Yesudas became a friend and colleague, and helped me to overcome language barriers and other practical matters. In Switzerland, Giorgio Semenza, who I consider extended family, helped arrange interviews and provided a sounding board for my often critical views of the subject. In Thailand, both Supat Attathom and Vilai Prasartsee took extensive time out of their busy work schedules to provide me with incredible hospitality during my visits.

Back at home, friends and family provided a backbone of personal support and inspiration for me during the writing of this book. My father and my late mother, who passed away during the writing of this book, deserve credit for raising me to be a critical thinker, ill content to leave any stone unturned. I am also thankful for their encouragement of my aspiration to become a scientist at a time when women were still underrepresented in the laboratory. When the strain of travel, research, writing, and raising a small daughter on my own appeared to be too much during the birth of this book, my sister Jill was always there for me.

A cadre of friends too numerous to name here cheered me on through the inevitable ups and downs that such a project necessarily exerts. I am truly thankful for my community here in Portland and for the sustenance, both literal and figurative, that life in city lush with fellow "foodies" offers.

On the practical side, I was supported by a number of literary professionals who helped see to it that this book made it into readers' hands. Jeanne Fredericks, my literary agent, persistently and enthusiastically presented my book to publishers, and offered steady guidance and encouragement. Stan Wakefield, editor at AMACOM, signed on the book with eagerness and excitement, even when the topic was ahead of its time. I am grateful to the staff at AMACOM for seeing this book through in a timely and professional manner.

I am also thankful for the generous support of a CAREER Award from the National Science Foundation. This material is based upon work supported by the National Science Foundation under Grant No. 0093827. Any opinions, findings, and conclusions or recommendations expressed in this material are those of the author and do not necessarily reflect the views of the National Science Foundation.

Finally, the person to whom I owe the most gratitude of all is my daughter, Kiran, who joined me in the course of my research travels for this book. The reward of having her in my life goes beyond words, and I am forever indebted to the patience and support that she offered, in her own tiny way, as this book took shape and form.

Food Fight

The Historical Roots of the GM Food Debate

"U.S. Ready to Declare War Over GM Food," threatened the headline that peered out from between the pink-tinged pages of London's *Financial Times*. It was January 2003 and the past six months had seen a vicious food fight take place on both sides of the Atlantic, with an unanticipated outcome. In the fall of 2002, six African countries had turned up their noses at food aid provided by the United States, suspicious of its genetically modified (GM) nature. Ultimately, five of the six nations had been cajoled into accepting the food aid, with some insisting that it be milled before distribution to prevent it from germinating and taking root to contaminate their fields. But Zambia—a small country centrally nestled in the heart of southern Africa—had held out and stubbornly said no thanks, even as U.S. officials condemned the rejection and accused Zambian leaders of crimes against humanity for allowing their citizens to starve.

By early 2003, focus on the African food aid crisis had died down, at least temporarily, and Zambia had found food from other nations to feed the country's hungry. But the Americans were still smarting from the rejection, and the administration of President George W. Bush now turned its ire on Europe, focusing on consumer opposition to genetically modified foods there, which had led the European Union to instigate an informal moratorium on approvals of new genetically modified imports. In response to Europe's insurgency over the GM food issue, Robert Zoellick, then the U.S. Trade Representative, had publicly referred to Europeans as "Luddites" and "immoral" and had accused them of steering starving Africans away from genetically modified foods.

But the transatlantic battle over GM food wasn't the only war brewing at the dawn of 2003. Since the previous summer, at about the same time that it had proposed sending humanitarian GM food aid to African countries, the Bush administration had begun seriously plotting the U.S. invasion of Iraq. And for the second time that summer, a proverbial David had stood up to an unsuspecting Goliath. Just as a swath of starving African nations were balking at accepting U.S. food aid, leaders and the proletariat alike in several European nations had delivered an emphatic "no" to backing the Bush administration's plan for war in Iraq. Historic demonstrations opposing the invasion were held in cities around the world, as deep divisions were cast between former allies in a Cold War that had long since melted away. New global battle lines were being drawn, and suddenly the dominance of the United States as a superpower was being challenged not only on the food front but on literal battle lines as well.

Onward, Freedom Fries

With European opposition to war in Iraq mounting in the early months of 2003, the Bush administration backed away from its threats to simultaneously pursue combat over the GM food issue. "There is no point in testing Europeans on food while they are being tested on Iraq," a senior White House official reported to the *New York Times* in February of that year. Nonetheless, food remained an important testing ground for U.S./European relations over the war in Iraq. Later in February, Cubbie's—a small, family-friendly restaurant in Beaufort, North Carolina, just down the street from the Cherry Point military base—hung a sign in its window announcing a political change of menu. "Because of Cubbie's support for our troops, we no longer serve french fries. We now serve freedom fries," the sign declared, as customers stepped up for patriotic second helpings.

In the increasingly fervent atmosphere surrounding the war, it took no time at all for this food fad to spread up the Atlantic coastline and into the cafeterias serving the U.S. House of Representatives. At a press conference in March 2003, two Republican lawmakers—one representing Cubbie's district in North Carolina—officially took the French out of the fries served to lawmakers in the nation's capital as an act of gastronomic patriotism. From there on out, "freedom fries" and "freedom toast" were the new haute cuisine. In New Jersey, French wines were let like blood when a restaurant owner ceremoniously decanted bottles of Merlot and Dom Perignon into a toilet while news cameras rolled. "I have no animosity toward French people, their culture, or their cuisine," he said. "I'm just trying to show my displeasure with their political policy."

Not everyone found the expression of patriotism through food and drink to their taste, however. A New York legislator worried that Belgian waffles and Russian dressing might need to go as well if those countries failed to back the war. The same legislator also wondered whether the fate of Mexican restaurants nationwide could be threatened pending Mexico's support. This was not the first time a nation's victuals had been held hostage in the name of war. While the current food front merely grazed the surface of the real political upheaval at hand, it harkened back to earlier protests during World War I, when frankfurters with sauerkraut were banned in favor of the newly christened "hot dogs" served with "liberty cabbage." In fact, the United States had practically been born with a political palate for food and war, with the Boston Tea Party serving as the christening to the revolution that founded the nation.

The Bush Administration Declares
War on Resistance to GM Food

By the time the Bush administration declared the supposed end to the official war in Iraq in early May 2003, the battle over naming the nation's fast
food was in retreat. But a new war over GM foods was just beginning. Two
weeks after the end of the Iraq conflict was announced, U.S. Trade Representative Robert Zoellick and then Agriculture Secretary Ann Veneman held a
joint press conference to announce that they were taking action against Europe through the World Trade Organization. The official press release documenting the filing of the WTO case was subtitled "EU's Illegal, Non-Science
based Moratorium Harmful to Agriculture and the Developing World." To
underscore their point, representatives from a number of developing nations
were paraded at the press conference. Each testified in his or her own ac-
cented English to the importance and salvation possible with the use of GM
crops and to the detrimental impact of Europe's rejection of this technology
in his or her homeland.

At the press conference, Zoellick had trotted out the oft-repeated
claim that Americans had been relishing GM foods for years with nary a
health effect in sight. He even embellished the claim this time with his
own smug observation that "Many Europeans must agree because, on my
flight home from Europe last week, I certainly didn't notice European tour-
ists coming to America lugging suitcases stuffed with food." But Americans hadn't always relished GM foods, and polls indicated that many
weren't even aware that they were eating them, despite the presence of
genetically modified ingredients in more than two-thirds of processed
foods on grocery store shelves. And European opposition to GM food ran
deeper than just politics or trade issues. Opposition to GM food hadn't
really started in Europe, either.

The seeds of resistance to GM foods hadn't actually originated in Europe as the Bush administration implied. In fact, they had been sown in the land of their birth, in the fertile soil of burgeoning U.S. biotechnology laboratories, albeit many decades earlier. The opposition to GM food that was now on the global battlefield had its roots in the vehement debates that had surrounded the first wave of genetic engineering technology. Genetic engineering had arisen in the 1970s, when scientists realized that the powerful tools they had developed might allow them to transfer DNA (deoxyribonucleic acid, the hereditary material found in humans and almost all other organisms) from one species to another, with unpredictable effects.

These early proposed experiments involved the splicing of DNA from a tumor virus into *E. coli*, a common resident of the human gut—a fact that had made many scientists worried. After a heated debate at a scientific meeting in 1971, where the idea for the experiments was first presented, such research was temporarily postponed. But the genie was out of the bottle, and within two years, scientists had struck upon techniques that allowed quick and easy cross-species mixing and matching of DNA in a far more promiscuous manner than the originally proposed viral method. By commandeering enzyme pathways present in a wide range of bacterial strains, scientists were now able to precisely snip out regions of DNA from one species and snap them together with genetic fragments from a totally unrelated species.

Added to the ongoing debate over the potential safety concerns of transgenic experiments involving tumor viruses, these new developments in recombinant DNA technology spurred furious debate among scientists. Maxine Singer and Dieter Soll—two prominent molecular biologists and co-organizers of the scientific meeting at which this new technology was brashly unveiled—felt compelled to write an official letter to the National Academy of Sciences. Singer and Soll warned that by using this technology, "new kinds of hybrid plasmids and viruses, with biological activity of unpredictable nature, may eventually be created . . . certain such hybrid molecules may prove hazardous to laboratory workers and to the public." The letter concluded with the recommendation that the academy establish a task force to investigate the dangers and develop regulatory guidelines governing this uncharted scientific territory.

Despite these early concerns of scientists involved in the birth of genetic engineering research, the public was not yet privy to such specters. Media reports of the new technology conveyed genetic engineering as a "marvelous achievement of science, rather than as dangerous, irresponsible tinkering with genes." It took an ominous letter from the National Academy of Sciences task force—made up of some of the most prominent molecular biologists in the field and calling for a moratorium on several types of recombinant DNA research—to get the notice of the press. Suddenly, newspapers across the United States exploded with the news of the potential dangers of biotechnology and the need to halt such research in its tracks.

Scientific Self-Rule at Asilomar

On the other side of the Atlantic, European scientists and research organizations took notice of the U.S. debate. While many countries reacted with concern, the range of views on the continent remained broad, falling along a wide spectrum of responses from an outright ban to no regulation at all. Meanwhile, in the United States, the issue was reaching crisis proportions. In 1975, sensing the need for intervention, the National Institutes of Health (NIH) and the National Science Foundation convened an International Conference on Recombinant DNA Molecules, bringing approximately 140 scientists to the sunny shores of Monterey Bay in California to discuss the dangers and develop regulatory policing guidelines for themselves.

The meeting came to be known as the Asilomar Conference after the rustic, camp-like conference center where it was held. The conference produced a flurry of discussion and debate surrounding the ethics and risks of the new genetic technologies. In a jam-packed three-day schedule, scientists discussed their options and butted heads over how the new science should be regulated and by whom. But in general, the converters were speaking to the converted, and the deeper ethical concerns that would later erupt in public debates never made it to the table. One panel description advised that it would "touch on ethical issues insofar as responsibility is concerned, and focus on the several components involved in balancing the risks and benefits of creating recombinant DNA molecules." While the issue had provoked anxieties, there was nothing particularly earth-shattering about the way in which it was being addressed here.

Although it may have led to few new conclusions, what was unique about this situation was that scientists were in essence regulating themselves. The National Academy of Sciences task force that had recommended the moratorium had been made up of the very scientists who had pioneered the snip-and-glue type of transgenic DNA technologies under debate. And the Asilomar Conference, though funded by the U.S. government, was a meeting of the minds that made science, not regulated it.

Such a situation created a vicious circle. As pointed out in an open letter directed to the Asilomar Conference from a group of concerned scientists calling themselves Science for the People, "this is like asking the tobacco industry to limit the manufacture of cigarettes." Not that those in attendance were unified in their recommendations, anyway. In the final session of the conference, negotiations broke down between those in favor of a tiered series of regulatory criteria and a small group of renegades (including the Nobel laureate James Watson) who voted to scrap the whole regulatory project and forge ahead unimpeded.

The end result of the Asilomar Conference was a relaxing of some of the original genetic engineering moratorium bans in favor of clearly delineated guidelines governing specific types of research. In due time, the NIH —the major government funding body for much recombinant DNA research—adopted these guidelines for all of its federally funded projects.

However, compliance remained voluntary, setting in place a laissez-faire regulatory scheme that would only worsen with political age and come back to bite with renewed ferocity in the GM food debate.

Despite the dangers of scientific self-rule, the Asilomar event represented a turning point, a moment of scientific self-reflection rare in a world where competition for the next Nobel discovery or front-page finding tended to put blinders on scientists' concerns for social and ethical ramifications of scientific research. Nonetheless, decades later, when regulations on the release of genetically modified food came to pass, the idea that those who made the science knew best would be brought forth once more. The idea would only gather momentum once such science was coupled with big business interests.

A Public Showdown in Cambridge

Following the Asilomar Conference, the recombinant DNA debate now seemed to have been concluded for many within the scientific community: The hazards had been discussed, regulations had been laid down, and finally the work could proceed. But the public was just beginning to catch on to the dangers that scientists thought they now had under lock and key. Harvard University's announcement in the spring of 1976 that it intended to build a new set of containment labs to carry out the controversial cross-species DNA experiments finally brought the first round of public opposition to biotechnology to the fore.

Harvard's Biological Laboratories building lies nestled toward the back of a small courtyard, guarded by two giant bronze rhinoceroses at the end of Divinity Avenue in Cambridge, Massachusetts. In the mid-1970s, a new crop of ambitious young professors held court in the building, side by side with esteemed colleagues whose research projects had inhabited the Biological Laboratories for years. For some of the more junior professors, the prospect of a high-security containment laboratory meant unbounded opportunities to explore the further reaches of the molecular underpinnings of living beings, to unravel and reweave the strands of DNA that controlled life itself. For others—like Nobel laureate George Wald and biologist Ruth Hubbard, whose laboratories abutted the proposed new containment facility—the proposal was an outrage, imposing risks on par with or even worse than those from nuclear radiation.

It was not long before Harvard's plans were discovered by a local tabloid newspaper, the Boston Phoenix, and made their way to the desk of Cambridge's charismatic mayor, Al Vellucci. Vellucci lorded over a city rife with town-gown tensions. Cambridge was home to both the prestigious Ivy League domes of Harvard and the techno-industrial laboratory compounds of MIT, institutions plying their intellectual craft in the midst of working-class, urban neighborhoods. Vellucci's last run-in with Harvard's administration had been a fabled spat over parking spot shortages presumably caused by out-of-state students and their expensive cars. The problem ended when Vellucci proposed paving over Harvard Yard to provide additional parking, and Harvard officials quickly found another solution. And now Harvard was at it again, planning a dangerous, high-risk project that could put the safety of Vellucci's jurisdiction in jeopardy. And he had to learn of it from an underground tabloid!

And so it was that Mayor Vellucci and his Ivy League rivals met once again on the council floor, ready to duke out the dangers and dreams stemming from recombinant DNA research. But this time, the team was split. Harvard's Wald and Hubbard—along with Jonathan King, a molecular biologist from MIT—supported Vellucci with testimony in favor of a city-wide moratorium on the type of research proposed for the new laboratories. Facing off against them was an equally prominent team, which included Mark Ptashne (dubbed "Comrade Mark" for his liberal political stance on Vietnam), the scientist in charge of the new containment facility. On Ptashne's side was Maxine Singer, one of the original organizers of the scientific conferences on recombinant DNA safety. Now convinced of the sufficiency of government oversight, she flew up from Washington to testify in favor of allowing the techniques to proceed.

In the end, the city council voted to set up the Cambridge Experimentation Review Board, a public oversight body charged with helping to develop guidelines to regulate recombinant DNA research at Harvard and MIT. But the council went further, implementing a temporary moratorium on the kind of research proposed to take place in the laboratory, and setting a precedent for local overrides of federal guidelines. Little did they dream that the same model would be implemented again decades later in a similar use of the science, when local governments from Vermont to California would pass moratoria banning the planting or sale of genetically modified seeds.

Risks and Benefits, Science or Science Fiction?

The Cambridge debates had injected a degree of fervor into the public deliberations over the emerging science of genetic engineering and recombinant DNA technology, and they had led to wild speculations regarding both the risks and benefits of the new technology. For example, there was rampant fear in some quarters that genetic engineering would increase the incidence of cancer. However, David Baltimore, a Nobel laureate famous for his work with viruses, argued that the technology would contribute to overcoming cancer rather than cause it. And James Watson—one of the discoverers of the structure of DNA—claimed that "The occasional rare exposure to a bacterial-derived 'cancer gene' should have negligible impact compared to the massive assaults we receive every time we are infected with one of the many, many viruses with oncogenic potential." In other words, there was far more danger of contracting cancer from day-to-day exposure than from genetic engineering.

At the same time, doomsday scenarios were invented in which an esteemed Nobel scientist would misuse his or her power. Perhaps he or she would proclaim in an acceptance speech: "All of the nations now possessing nuclear weapons will totally dismantle and destroy them within six months. If not, then my lysogenic versions of the common rhinoviruses will make it fatal to contract a common cold." And in a hypothetical scenario that linked the technology to racial tensions (and inversely foreshadowed elements of future conspiracy rumors regarding the HIV virus), the head of an African state was imagined to proclaim: "Past threats to our sovereignty made necessary the secret development of a deterrent 'ethnic weapon' germ which, if released, would cause eventual cancer in 96 percent of Caucasians. Now, we must apologize, it appears that a laboratory worker's error has resulted in the irreversible spread of this microbe among the local monkeys."

The Biotech Boom

In the end, though, it appeared that the benefits won out. By the 1980s, biotech was booming. The year 1982 saw the first Food and Drug Administration (FDA) approval of a biotech drug: Humulin, a genetically engineered form of insulin that would go on to win Genentech, its maker, kudos on the cover of *Business Week* as "Biotech's first superstar." Herbert Boyer, Genentech's cofounder, had earlier graced the cover of *Time*, depicted cartoon-like as a smiling, chubby cherub floating above electron micrographic strands of DNA hovering next to the headline "Shaping Life in the Lab: The Boom in Biotech."

In Cambridge, Harvard had built not just a new laboratory but an entire new building dedicated to the types of experiments that had been under debate. A fleet of late-model Porsches began to turn up in the parking lot, as the scientists whom Mayor Vellucci had tried to stop became major players in the burgeoning biotech industry that was injecting new economic strength into the city and region. Who could object to that? Besides, it was

the 1980s and the political tide was turning conservative, as the radical protest era of post-Vietnam gave way to the "me" generation.

And biotech was producing results, results that helped people. Following the success of Genentech's recombinant insulin, the FDA in 1986 approved the world's first genetically engineered vaccine, aimed at protecting
against Hepatitis B. A year earlier, genetic fingerprinting had entered the
courtroom, and the NIH had put in place guidelines to govern human gene
therapy. It appeared that the cures had prevailed and the fears had been
foiled, since there were no reports of escaped *E. coli* swimming up out of
sewers, and the tools of molecular biology were turned toward understanding and fighting a much more serious new plague: AIDS and the virus that
causes it, HIV.

In contrast to the anxiety over uncontrollable outbreaks and bioterror
that recombinant DNA had spawned in the 1970s, the biotech boom in the
1980s gave rise to hope and optimism, as investors rushed to fund this
new brainchild business and its anticipated bounties. When Genentech went
public in 1980, it sparked a stock market revolution, soaring with the most
spectacular ascent in value that Wall Street had ever seen. When the company received regulatory approval for a biotech heart drug later that decade,
it celebrated with the kind of hubris that only a young industry would be
bold enough to proffer: champagne under circus tents, a full-on rock and
roll concert, and fireworks so bright that they shut down San Francisco International Airport.

The Bubble Bursts

Then the biotech bubble started to burst, as quick fixes and miracle cures
failed to materialize as quickly as speculators had hoped and the U.S. economy began to weaken. All was silent on the genetic engineering front for a
time. By 1992, the biotech stock market had sunk into a steep descent, and

smaller start-ups were gobbled up by mega multinationals. Biotech took on a new face—kinder, gentler, and focused more on the traditional big pharmaceutical industries that carried the experience and especially the cash flow to see their developments through costly, high-risk clinical trials, approval, and marketing.

But that was just on the surface. The public concerns over the risks and unknowns inherent in recombinant DNA technology that had arisen in the 1970s had not been mollified. They had merely been swept under the rug, to proliferate and await a more opportunistic moment. Throughout the 1980s, the indefatigable activist Jeremy Rifkin kept the fires of opposition burning, even if they were for the most part mere embers on the larger social and political smokescreen, hidden for the moment behind the do-good aura of blockbuster biotech drugs. The deeper, more insidious side of genetic engineering had gone underground, but occasional outbreaks of angst still turned up every now and then.

During the 1980s and into the early 1990s, small oppositional blips continued to appear on the biotech radar detector. The coup de grace was perhaps the "ice-minus fiasco" in 1987, in which Rifkin successfully sued the government to halt the release of genetically engineered bacteria designed to prevent the formation of ice crystals on crops—in this case, strawberries. Rivkin's group argued that such bacteria could wreak havoc with weather patterns if they multiplied and escaped into the upper atmosphere, turning snow into rain or even leading to drought. The group in effect was able to put a temporary government moratorium on the first field release of a genetically modified organism. In the meantime, it was revealed that the company producing the bacteria had already illegally released it among the trees on its rooftop garden.

When approval for field testing was finally granted and company representatives traipsed into the appointed strawberry fields—trailed by government officials and a herd of media reporters—they found quite a surprise, one that was strikingly similar to the scenario that would crop up in the fields of Europe some ten or twenty years later. Nearly all of the strawberry plants had been pulled up and placed limply on the ground, where they lay withering in the sun. Company officials rushed to resuscitate the plants, propping them droopily back up in the ground. The officials also donned space-age style "ice-busters" safety suits, as mandated by the Environmental Protection Agency (EPA), in what had become merely a ceremonious but now vacuous public relations stunt.

Ice-minus wasn't the only genetically modified crop application in the making during biotech's first golden age. Scientists and industry had tried hard to put a decidedly more palatable, health-focused face on the future of biotech in their public prognostications during the early years of genetic engineering. But locked away in a world of trade secrets and high-stakes competition, the future of GM food was busy germinating in the small start-up ventures as well as the goliaths of the agricultural industry. Such research was not limited to the United States. The tinkering had been taking place in Europe as well, as multinational corporations and smaller research-driven enterprises alike sought novel ways to shuffle the genes of agricultural crops to improve productivity and most of all increase profit. Although it would take a few more years before genetically modified foods turned up on their dinner plates, through the early 1990s, the general public on both sides of the Atlantic remained largely unaware of the agricultural changes that were brewing in some of the nation's leading seed laboratories. It would not take long, however, for them to find out.

CHAPTER 2

Out of the Swamps and Up from the Soil

DNA at the Dinner Table

Back in the 1970s, as scientists were putting DNA from different species on the chopping block and cooking up new transgenic recipes, a synthetic amino acid called glyphosate had quietly made its way through the U.S. patent system. Most people know amino acids as essential dietary components, the building blocks that make up proteins. But glyphosate is not the kind of amino acid that you want to eat, at least not in large doses. Rather than helping to build up proteins, glyphosate actually blocks the synthesis of essential amino acids in plant cells, leading to their death. This is exactly what makes it a powerful herbicide, and this is what motivated the quest to patent its biochemical structure. In the form of the herbicide Roundup, glyphosate sits at the nexus of Monsanto's domination of the GM seed industry. It has now also become a centerpiece in the story of opposition to GM foods.

Because it degrades relatively rapidly when exposed to sunlight and heat, is cheap to manufacture, and has relatively low human and environmental toxicity, glyphosate in the form of Roundup became an instant winner for Monsanto in the 1970s. In an industry still struggling to emerge from its chemical ties to mid-twentieth century warfare production, Roundup offered a kinder, gentler type of product that could be marketed to the farmer and homeowner alike. Throughout the 1970s and into the 1980s, Roundup drove Monsanto's profits and earned itself a household name. But all good things must come to an end, and the standard life of a patent is only seventeen years. Once proprietary protection expires, a product like Roundup is likely to generate a feeding frenzy for competitors who can gorge on its reputation in the marketplace but make and sell it for less. Monsanto desperately needed to find a way to extend the proprietary life of its block-buster product.

Roundup's Second Life

The fact that the patent on Roundup would not outlive the end of the twenti-eth century was no doubt not lost on the powers that be at Monsanto. The company was dependent on Roundup to prop up profits as it struggled to reshape its identity from that of the Cold War chemical supplier it had once been into the forward-thinking biotech innovator it hoped to become. During the 1980s, Monsanto had annexed a rash of biotech firms in the arena of healthcare, but the acquisitions had failed and were eventually shed. And then the technology for Roundup Ready crops emerged. Scien-tists at Monsanto struck gold in a mine they almost didn't think to explore—bacteria slumbering in toxic waste swamps surrounding a manufacturing plant, in mud contaminated by sickly soups of chemicals, including Roundup.

Living systems are highly adaptable to numerous forms of stresses and strains. Evolutionary selection favors such adaptability in the form of mutations, which are alterations in genetic codes that help to create the variety on which all of life thrives. Without random mutations to mix things up a bit and provide a certain spice to life, organisms would have no way to adjust to the vagaries of unpredictable and ever changing environments, and life as we know it would soon die off.

And so it makes sense that a few lucky bacteria living at the bottom of a contaminated bog would be able to genetically adapt to living under a constant siege of chemicals, including the herbicide Roundup. By the mid-1980s, Monsanto's scientists had recognized that one way they could extend the life of Roundup—their poster child product that was fast approaching a difficult adolescence—was to piggyback other products onto its success. Genes that made crop plants resistant to Roundup would do just that. But how to make a plant resistant to the potent effects of an herbicide?

One way would be to genetically resupply the enzyme that Roundup destroyed. Monsanto's scientists had tried that, and while it worked in the laboratory, the plants still wasted away and died under the intensity of application found in field spraying. The scientists made numerous attempts to artificially reengineer the shape of Roundup's target protein, and while some came close, none of them were tough enough to withstand the level of commercial herbicide applications favored by U.S. farmers. Just when Monsanto was running out of ideas and competitors were threatening to overtake their efforts, scientists at Monsanto remembered that lurking in the backroom of their company's waste cleanup division lay a virtual cupboard of treasures.

Genetic Gold from a Vial of Mud

Monsanto does not manufacture its Roundup products in the vicinity of St. Louis, where its corporate headquarters are located, but in a small town called Luling, located along the Mississippi River in southern Louisiana. Luling is on the shores of what has come to be known as "cancer alley" for the massive amounts of toxic carcinogens—chemical and industrial pollutants—that spew onto the river's banks. Not surprisingly, high levels of Roundup are found in the swampy waste ponds surrounding Monsanto's manufacturing site. The Roundup exerts a constant and unabating selective pressure on the microbes living there to develop tolerance to the chemical assault. "Resist or die" is the harsh and unavoidable evolutionary message transmitted to the genes of anything attempting to survive in these polluted pools. And amazingly enough, they do survive. Through clever rearrangement of their genomes and just the luck of the mutagenic draw, colonies of resistant bacteria have found ways to thrive in the muck, modifying their enzymatic pathways to outwit Roundup's biochemical battle plan and neutralize its toxicity.

Years before, samples of these lucky bacteria had been brought to Monsanto's Creve Coeur campus outside St. Louis, as part of routine surveillance of the chemical sludge outside the Luling plant. But the genetic goldmine held within these tubes of mud had never been explored, at least not in the context of their ability to resist the effects of Roundup. After all, Monsanto did not want to pursue resistance to its miracle product, let alone publicize it. Roundup was known for its amazing effectiveness, its ability to yellow a field of any and all green, to act in a nonselective and preemergent fashion and then swiftly disappear from the scene. A genetic revolt that would allow plants to circumvent the effects of Roundup was not something that Monsanto saw as helpful to promoting its product.

That is, not until this point. Once Monsanto's scientists began to toy with the idea of making agricultural crop plants that were selectively resistant to Roundup, those tubes of mud began to look a lot more promising. The scientists knew that the genes they were looking for must be hidden within the mud. If they could isolate the magic snippets of DNA and then transfer them to crop plants, farmers would be free to douse their fields with as much Roundup as they liked—before, during, and after the growing season, wiping the fields clean of weeds while at the same time sparing the important cash crops. At present, Roundup could be applied only prior to planting, lest it kill off the desired sprouts once the crop began to grow. But if genes encoding resistance to Roundup were built into the seeds they planted, farmers would be able to spray Roundup again and again. This would keep the market for Roundup strong and stable.

Within a few years, Monsanto's scientists had ferreted out the resistant form of the gene, inserted it into crop plants, and struck success. They had made plants that were "Roundup Ready," poised to stand tall under even the harshest of glyphosate treatments. Since the company was not itself in the seed selling business, in 1992, Monsanto licensed its Roundup Ready technology to Pioneer Hi-Bred International, a seed purveyor that controlled one-fifth of U.S. seed sales and 40 percent of the corn market. The genie was out of the bottle.

Bt: An Organic Pesticide Gets a GM Makeover

Resistance to Roundup wasn't the only trick that Monsanto had up its sleeve. Since the early 1980s, scientists at the company had been toying with the genes from a small but powerful soil bacterium called *Bacillus thuriengensis*, or Bt for short. The scientists were trying to coax Bt's toxic properties out of its sporelike forms and into target crops.

Bt lives out its single-celled life beneath the surface of the soil, balling up spores when conditions turn harsh, and storing up inside them an impressive supply of insecticidal toxins to protect it during this hibernation of sorts. The same selective evolutionary forces that pressed Roundup resistance genes into action in the bacteria living in Monsanto's waste ponds have in nature also encouraged the expansion of a wide variety of toxin proteins in different species of Bt organisms, each specific to a small group of insect pests. Just as the Roundup Ready genes that Monsanto discovered in the waste pond bacteria cleverly neutralize the effects of Roundup, the toxin genes of the Bt organism are able to lethally outwit the insect predators that might decimate their population in the soil.

When the larval form of a hungry insect ingests a Bt spore, the alkaline conditions in its gut release the poisonous properties, paralyzing the intestine and blocking the absorption of nutrient uptake in a fatal version of something akin to caterpillar diarrhea. Generations of selection have molded different varieties of Bt toxin lethal to very specific insect targets. They have fine-tuned its deadly effect to act only at the predatory larval stage, making Bt toxin an effective and highly specific pesticide that also degrades relatively rapidly when exposed to sunlight.

Bacillus thuriengensis isn't merely a marvel of genetic engineering, however. It was first identified in 1901 in Japan, where it was implicated in a deadly outbreak that threatened to decimate the silkworm industry. The species was officially christened *Bacillus thuriengensis* in 1911 after the German province of Thuringia, where the bacteria was found fending off the flour moth in bags of stored grain. Commercial preparations of Bt spore sprays have been used by organic farmers as protection against a variety of moth and butterfly family pests since the late 1950s.

In 1962, Rachel Carson—doyenne of the environmental movement and foe of the chemical pesticide industry—wrote effusively in praise of Bt in her classic book *Silent Spring*. She pointed to Bt as an example of a sustainable alternative to the synthetic chemical pesticides wreaking havoc on the world's ecosystems. Carson extolled the successful trials of Bt as a pesticide to protect a broad spectrum of plant life from insect infestation, from bananas invaded by root borers in Panama to the coniferous forests plagued by gypsy moths in North America. Carson offhandedly lamented that while for most crops, an external dusting of Bt spores would suffice, for the tall evergreens in the northern United States and Canada, "The main technical problem now is to find a carrying solution that will stick the bacterial spores to the needles of the evergreen."

Most likely, Carson would never have imagined the solution that Monsanto scientists would eventually capitalize upon to address that very problem. Nor would she likely have approved of it, either. Although Monsanto wasn't much interested in applying Bt to evergreen trees plagued by gypsy moths, the company was concerned with the same issue that Carson had pointed out: how to get Bt on a plant and keep it there. Bt sprays were effective and approved for use in organic agriculture because of their short half-lives and relatively low toxicity; within a fortnight following spraying, Bt would be degraded by exposure to the elements. For a company bent on profit and patent protection, this was not seen as an advantage. Scientists at Monsanto were after a way to capture the toxic properties of Bt and encode them within seeds, which they could then proprietarily sell as pesticide and plant in one.

Things did not exactly go as simply as planned with the quest for Bt, however. Unlike most other plant genes that scientists had worked with, the crystal toxin from Bt was difficult to regulate once it had been inserted into a plant's genome. Eventually, with the right combination of on and off switches inserted in front of it, the Bt gene seemed to churn out enough copies to kill off measly laboratory-raised caterpillars. But it was still not enough to face the nemeses of pests in the field. For this, scientists needed to conquer the transgenic divide between bacteria and higher organisms.

It has been common knowledge for nearly fifty years that all living organisms on earth share a common genetic code. That code is stored in the base pairs of DNA, transcribed into a messenger form of RNA, and then finally translated into the amino acids from which proteins are strung together. The discovery of unique phenomena—such as retroviruses like HIV, which reverse the DNA-RNA steps, and prions, the causative agents of mad cow disease, which encode and transmit information directly at the protein level without any DNA at all—have to some extent challenged this central dogma. But for the most part, all of life steps in time to the DNA-RNA-protein beat. However, when it comes to the translation step, bacteria and more complex life forms march to their own tunes, or at least sing those tunes in different musical scales.

It seems that although the genetic code remains consistent among all species of life, bacteria have a distinct preference for a subset of amino acid codes, one that is less familiar to higher organisms such as plants. Thus, when the amino acid code from a bacterial gene is called into action in a plant, the "pitch" and "frequency" need to be adjusted, in order for the protein to be manufactured most efficiently. This was precisely the challenge that faced scientists attempting to coax plants into spreading Bt toxicity to insect pests in the field. The entire genetic score would have to be rewritten to harmonize with a plant's genetic translation scale, which was no small undertaking.

Ultimately, the biotechnology industry managed to perfect the production of the Bt crystal toxin in a wide variety of agricultural plants, setting the stage for mass marketing of pesticidal plants. But while plants containing genes for Bt toxicity and glyphosate resistance—the future autocrats of an emerging agricultural biotechnology industry—were still in the making, another GM product beat them to market.

The Search for Year-Round Tomatoes

The Flavr Savr tomato was a perhaps unlikely victor for first prize in the GM food approval race. This genetically modified tomato was the brainchild of Calgene, a small, private start-up company based in California. Calgene, recognizing the power of moving a plant's DNA around, had been tinkering with a number of genetically modified food and agriculture products since its founding in 1980. While Monsanto and other "gene giants" built upon a base of multinational success in the chemical or pharmaceutical industries, Calgene relied upon a small, dedicated, and hardworking band of renegade scientists toiling out of Davis, California—the heartland of U.S. agricultural innovation. It was here that the novel idea was launched of halting a gene involved in tomato ripening. This utopian goal would yield a tomato that could withstand long hours of transcontinental transport without paying the price of the pallid flavor and winter mushiness of tomatoes that had been picked green on the vine.

What could be more appealing than a ripe, red tomato, bursting with flavor in the middle of the winter, but hardy enough to be shipped cross-country? Given that the average food product moves 1,300 miles to reach its grocery store consumer, the dream of engineering a tomato that could ripen naturally yet still retain its resistance to the bumps and bruises of a long road trip seemed highly alluring.

Rewriting a Tomato's Recipe for Ripening

Tomatoes ripen partially through the action of a culprit gene that encodes an enzyme called polygalacturonase, or PG for short, which leads ripe fruits to depolymerize their pectin, softening the fruit's flesh through the breakdown of its cell walls. Calgene's task was to find a way to slow down that gene, thereby preventing the ripening tomato from softening yet allowing it to retain its natural flavor and color.

When a cell needs to produce a protein, such as PG, it looks to its recipe book inside the nucleus: the string of DNA base pairs known as its genome. Encoded within these strands are the recipes for each and every protein the cell needs to produce, at any point in its lifetime. Recipes are written every which way along the strands, occasionally overlapping, some written in one direction, others running opposingly along the complementary strand, as if the genome were saving paper by writing its formulas on both sides of the sheet. But there are occasional blank pages inserted as well, long stretches with nary a recipe nor even a decipherable note written within its margins. These regions of so-called "non-coding" DNA sequences are hypothesized to provide structural or logistical support for the information bank, as if to bulk up the cookbook and make its spine more sturdy than it otherwise would be.

When an elaborate series of seasonal cues tells a tomato that it is time to ripen, the cells that make up that tomato thumb through their genetic recipe book and order up a batch of the PG enzyme. The recipe for the PG enzyme is copied many times over into a snippet of complementary RNA, then exported from the nucleus into the cell's cytoplasm, where it can be interpreted and acted upon by structures called ribosomes. These cellular chefs of sorts gather together the amino acids (the raw ingredients of protein manufacture) and line them up in just the right order to prepare a freshly active batch of PG enzyme for the cell to use.

In the case of the tomato, a fresh batch of PG enzyme will catalyze the breakdown of the pectin in the cell walls, leading to a softer, juicier tomato, in contrast to the hard and mushy texture usually associated with grocery store tomatoes that are reddened artificially rather than left to ripen on the vine. Other genes and processes are responsible for the enhanced flavor and color associated with ripe summer tomatoes, so that striking out the action of PG would be expected to counteract only the texture issue—an important factor in transportability of tomatoes, leaving the ripe flavor and summery aroma intact.

Calgene's Success—and Failure

Within a few years, scientists at Calgene had tapped into a novel way to inhibit the action of the PG enzyme without affecting other aspects of to-mato ripening. Since at the time the market for fresh tomatoes in the United States was topping $4 billion per year, this was quickly embraced as a poten-tial goldmine for the company. If Calgene could swiftly and cleverly erase the action of the PG gene, it could produce tomatoes that would have all the flavor and color of a fresh vine-ripened fruit but be strong enough to with-stand their long miles of transcontinental transport. Consumers would no longer have to put up with tomatoes that were picked green and then gassed with ethylene, a ripening agent, once they arrived at their destination, turning the tomatoes red but failing to impart flavor and texture. What was more, busy families would be able to stock up on tomatoes the same way they were beginning to hoard items like toilet paper and frozen chicken wings, buying in bulk and storing up supplies in their pantries. Flavr Savr tomatoes—as the PG-less tomatoes were soon christened—would keep not only their flavor but also their plump round firmness for up to three weeks in the refrigerator.

The technology seemed promising. As in the case of the Bt and glyphosate resistance genes, all the Calgene scientists needed to do was fiddle around with the regulatory mechanism of the PG gene, turning it down until its volume was barely audible. This would spare some of the pectin in the cell walls and keep the fruit firm and plump. But in this case, Calgene's scientists decided to try a novel approach to PG gene regulation, tinkering not with its on/off switch but attempting to blind the cell's ribosome from reading the recipe for the PG protein once it was transcribed, in a procedure known as "antisense" technology.

This approach seemed to work well, at least in the laboratory; PG protein levels were dramatically reduced in all parts of the tomato plant. But when those same tomatoes were planted in field conditions outdoors, surprisingly, they ripened the same way that normal tomatoes did. They did hold out better the closer they came to rotting. Thus, once the ripe tomato had been shipped, purchased, and taken home, it could sit it out in the fruit bowl a good week or so longer than a conventional tomato, which would rot much faster at that point. But along the way, the tomato had the same delicate squishiness that a normally ripened tomato would, compared with the golf-ball hard consistency of green tomatoes, which made the latter the choice on the long-haul shipping ventures required of the twenty-first-century food marketplace.

This was not good news for a company that wanted to produce a year-round tomato that seemed summer-fresh. Certainly, people would be willing to pay a premium for that fresh-from-the-vine tomato flavor in the dead of winter. But having to ship them using white-glove service would add prohibitory expense, potentially negating the profitability of the Flavr Savr trait.

When the first transcontinental truckloads of Flavr Savr tomatoes arrived at their destination, it was worse than anyone could have imagined. There were virtually no salvageable tomatoes on board. The staff had to use snow shovels to clear the debris. It was clear that transportability was going to be a problem.

Ultimately, the enhanced cost of shipping the Flavr Savr tomatoes, along with a few unfortunate coincidences in weather and other unplanned growing mishaps, spelled their death in U.S. markets. Despite consumer interest and widespread market availability, the company continued to struggle with its profit margins, delivering a product that cost more to ship but held longer on the shelf once it was there. Calgene referred to this as a "back end" benefit (perhaps not the most commercially savvy of euphemisms). The market didn't seem willing to pay enough for a back end benefit to make it worth the while. By 1996, Calgene had been gobbled up by Monsanto, and the Flavr Savr's shelf-life was beginning to expire.

Ironically, it was only in England—soon to be the seat of heated antibiotech food fervor—that the Flavr Savr tomato met with at least some commercial viability. There, Zeneca Plant Sciences licensed the technology from Calgene and used it in a strain of plants geared toward the processed tomato paste market. Reduction of PG levels in tomatoes that were meant for canning allowed processors to eliminate a costly pectin deactivation step and pass on that cost savings to consumers. Since the tomatoes weren't transported fresh to the shelves, the delicate handling required to bring the Flavr Savr fresh to supermarkets was no longer an issue.

Beginning in 1996, cans of genetically modified tomato paste began appearing on the shelves of British supermarket chains like Sainsbury's and Safeway. Consumers gobbled the new product up, all but ignoring the "made from genetically modified tomatoes" label in favor of the reduction in price. But this success did not last long. At the same time that shoppers were embracing GM tomato paste, another more ominous threat was on the horizon. That threat was not directly related to genetic engineering, but it would eventually spell doom for the further acceptance of genetically modified foods in England and indeed across all of Europe.

In the Shadow of Mad Cow Disease

Perhaps one of the reasons that the Flavr Savr tomato had escaped consumer opposition in England was that there were bigger food controversies brewing there at the time. In the early 1990s, Britain was so caught up in defending its beef that it barely noticed a lowly tomato can on the shelves. Mad cow disease, or Bovine Spongiform Encephalopathy (BSE), had struck the British Isles with a fury, leaving the British beef industry as reeling and disoriented as its cows. First France and then other European countries imposed a ban on British beef, even as the British government assured consumers that it was safe to eat and that the devastating brain wasting seen in cows could not be transmitted to humans. Reigniting a latent and longstanding feud between France and Britain, British grocery stores began displaying Union Jack-stamped beef, advertising its safety and superiority and urging a sense of nationalism in promoting consumer purchase and pride in its beef.

Then, in 1996—just as Flavr Savr tomatoes were silently slipping into oblivion in the United States and immigrating in their labeled cans into the U.K.—the fatal link between mad cow disease and human Creutzfeldt-Jacob Disease came to light. The government regulators' assurances had turned out to be hollow after all, and there was an apparently transmissible human form of mad cow disease. BSE could be contracted by eating infected beef. Consumers first held their breath and then gasped as the number of human cases mounted.

Opposition Erupts in Europe

It was into this emotionally heightened context that the ship bearing the first imports of Roundup Ready soybeans sailed into Europe. Waiting to escort it were activists from Greenpeace, who decried the shipment of genetically modified soy as a "floodgate to genetic pollution." They accused its importers of secretly attempting to mix GM ingredients into the public's food supply without their knowledge.

Across Europe, consumers began to condemn the use of genetic engineering in food products. Within a year, Greenpeace had solidified its forces protesting genetically modified organisms (GMOs) and blockading ships into an official "Genetic Hazard Patrol," complete with its own interception vessel christened *Sirius*, which was used to block the port of call of any ship suspected of bearing genetically modified cargo from the United States. By now, three European nations had erected legal barriers to GM food, and a growing number of grocery chains and public interest groups were decrying the lack of safety testing and potential environmental impacts of what they viewed as novel and unnecessary foodstuffs being imposed on them from across the Atlantic.

Into this fray jumped England's Prince Charles, who at the end of 1998 initiated an Internet discussion forum to ponder the ethics of GMOs. Not surprisingly, people leaped at the opportunity to make their voice heard on England's latest food controversy. More than 10,000 replies were posted, few of them favorable. By the following summer, Prince Charles had gathered enough of his own data to formally come out in opposition to genetically modified foods, which he did in an article published in the newspaper *The Daily Mail*. Genetically modified food, according to the heir to Britain's throne, was unnecessary, unproven, potentially unsafe to human health, and a threat to the environment. He countered the argument that genetically modified foods would solve world hunger—a message already being floated by the multinational purveyors of this suddenly scathed commodity—as an act of "emotional blackmail." Moreover, he denounced the development of GM foods as "the industrialization of life itself" and leading humanity toward an "Orwellian future."

Only a few years earlier, GM tomato paste had barely caused British consumers to blink. Now, however, neither they nor the supermarkets that sold to them wanted anything to do with this new technology. The biotechnology industry—seeing the writing on the wall as an important market sank with every ship blockaded at a European port—began to decry the "Luddites" and terror-mongers who they claimed were using scare tactics and anti-science rhetoric to frighten consumers away from their business-as-usual foodstuffs.

Frankenfood Furor

As the twentieth century drew to a close, the biotechnology industry recoiled as the term "Frankenfood" caught on like wildfire. The industry seemed to be surprised that people viewed GM as an unwanted dinner guest that had turned up at the table. One pro-biotech website defined a new syndrome that was frequently observed in Europe as "Frankenfood Frenzy," identifiable by the following traits: "Combine lots of emotionally-charged doomsday rhetoric with a good amount of anti-capitalist sentiment. Add just a pinch of scientific uncertainty about safety and you've created enough 'Frankenfood Frenzy' to serve the world."

But the term "Frankenfood" and the frenzy that surrounded it had not in fact emerged in Europe. Like GM food itself, the term had been imported from the United States, where it had first emerged in an editorial in the *New York Times* in 1992. The term had been ignored, left to fester and later inoculate susceptible European consumers.

In fact, Frankenstein had been at the table from the start, often seated side by side with the very scientists leading the genetic engineering revolution two decades back. Accompanying a September 1976 article entitled

"The Case Against Genetic Engineering," the New York Academy of Sciences magazine *The Sciences* ran a cartoon showing Frankenstein lurking in the background of a bastion of spooky chimeras created by an enthusiastic scientist. The following year, the *Bulletin of the Atomic Scientists* ran an article entitled "An Imaginary Monster" penned by James Watson, the co-discoverer of the structure of DNA. Even *Rolling Stone* magazine delved into the fear and uncertainty of the science of genetic engineering at that time, covering what it called "The Pandora's Box Congress" in the June 19, 1975 issue.

Yet all of these cultural warning signs had been quickly forgotten as scientists and industry—in their zeal and passion for a new and unknown technology—had failed to address their guest by name. They had attempted to brush the fears that references to Frankenstein represented nimbly off the table as if mere crumbs. But they were not crumbs, and as the mounting resistance to genetically engineered food at the dawn of the twenty-first century was to show, no amount of rhetoric regarding hunger, poverty, and doing good could make Frankenstein and his fearful foodstuffs go away.

CHAPTER 3

Of Politics and Precaution

"Frankenfood" Frenzy in Europe

As the 1990s wore on, opposition to GM food in Europe grew with incredible speed and vigor, fueled by mistrust in government regulators over the mad cow disease—or Bovine Spongiform Encephalopathy (BSE)—crisis and concerns over corporate commandeering of local food supplies and traditions. Most analysts of the anti-GM phenomenon in Europe would later cite 1998 and the furor over BSE as a major turning point in consumer attitudes toward genetically engineered food. But while the Flavr Savr tomato had narrowly escaped consumer rejection in the mid-1990s, a latent mistrust of biotechnology had been brewing behind the scenes in Europe long before the BSE crisis made the news.

As early as 1992, anti-biotechnology activists in Switzerland had managed to convince the public that biotechnology was something to be concerned about. Switzerland boasts a direct democracy form of government that allows citizens to introduce referenda to a national vote upon collection of 100,000 signatures. It is a wealthy and relatively conservative nation nestled in the center of the European Union (EU), yet it is not part of that common European economic and political community. Famous for the precision of its watches and trains, the purity of its chocolate, and the beauty of

31

its snow-covered Alps, Switzerland has few natural resources. Yet the nation boasts a highly educated and technically trained workforce, with more than its share of Nobel laureates per capita. Switzerland is also home to many multinational pharmaceutical and biotechnology companies, including Novartis, Roche, and Syngenta. Syngenta is one of the world's primary suppliers of genetically modified seed.

The Gene Protection Initiative

Despite signs that on the surface seemed to portend fertile ground for biotechnology to take root, beginning in 1992, Swiss anti-biotechnology activists began planning to put forth what was termed the Genschutz-Initiative, or Gene Protection Initiative. The final form of this wide-sweeping initiative proposed to forbid "the generation, purchase, or distribution of transgenic animals; the release of genetically altered organisms into the environment; and the patenting of transgenic animals and plants, of their components, and of the relevant processes." Moreover, imbedded in the initiative was a clause requiring that "experiments with all genetically modified organisms require proof of benefit and safety, proof of the lack of alternatives, and a statement of ethical responsibility."

By October 1994, supporters of the measure had gathered the requisite number of signatures needed to put the initiative to a vote. In the process, organizers had rallied the support of environmental groups such as Greenpeace, the World Wildlife Fund, and the Swiss Organic Farmers Association, as well as a number of women's and religious groups, including the Swiss Lutheran Women's League and the Swiss Catholic Women's League.

The consequences of passage of the Genschutz-Initiative would have been severe, impacting much of the scientific research taking place across the country. The consequences would certainly have had a profound effect not only on Switzerland's significant biotechnology and pharmaceutical industry but also on its basic scientists plying their trade in the country's many research universities. According to one scientist active in the debate, "Acceptance of the initiative would have ended Switzerland's preeminent role in biomedical research and prompted many biologists to leave the country."

Nonetheless, at the time that it was introduced, the Genschutz-Initiative had broad appeal, and polls showed that a majority of the nation's voters supported its quest to ban recombinant DNA research and the patenting of life. Seeing the writing on the wall as public support for the measure mounted in the years building up to the vote, Switzerland's scientists— usually a conservative, quiet lot who devoted their energy to their research laboratories rather than politics and the plebeian masses—turned out in force. On April 28, 1998—donning lab coats over their suits and ties—a band of Swiss scientists took to the streets, waving banners and staging a demonstration of their own opposing the restrictive Genschutz-Initiative that they saw as curtailing their livelihoods.

Ultimately, when it came to a vote in June 1998, the Genschutz-Initiative was soundly defeated. The public was largely swayed by practically motivated health arguments suggesting that passing the measure would curtail the development of lifesaving drugs and treatments. Backers of the measure—who were outspent approximately eight times by the pharmaceutical, food, and agrichemical industries— decried the emotional manipulation embedded in larger than life-size advertisements showing an elderly man backed by a flock of grandchildren begging viewers to "Say No to this initiative to save our grandpa."

In retrospect, anti-biotechnology activists realized that they had drawn with too broad a brush, and they were determined to focus their future efforts on GM food and the environmental consequences of field releases. Opponents of the measure breathed a sigh of relief but realized that this close brush with public scrutiny would not be their last.

Crafting the Cartagena Protocol

Wide-ranging anti-biotech legislation may have failed in Switzerland, but that did not mean that GM food was met with general acceptance. In the fall of 1998, the European Union quietly stopped approving new GM crops for import or cultivation on European soils. The following year, five EU states declared a de facto moratorium on GM crops until a regulatory system could be drawn up that would require detailed traceability of all GM products and their labeling on supermarket shelves. Momentum against GM food in Europe was building, and as individual EU member states continued to institute bans of their own under a "safeguard" clause of EU regulations, the GM door appeared to have been shut if not locked tight.

The Earth Summit Takes Up the GM Fight

Further strengthening the EU's mettle in standing up against the import or cultivation of GM crops was the impending Cartagena Protocol. The protocol was a global initiative stemming out of the United Nations Conference on Environment and Development, which had been held in Rio de Janeiro, Brazil, in 1992. At this so-called "'Earth Summit," representatives from 172 governments and more than 2,000 non-governmental organizations (NGOs) met to brainstorm around environmental sustainability in the face of ever encroaching globalization.

Several important documents were produced following the meeting, including the Convention on Biological Diversity and the Rio Declaration on Environment and Development, a short outline of twenty-seven core principles to guide international cooperation in "protecting the integrity of the global environmental and developmental system." Tucked into this latter document as guiding principle number 15 was a statement affirming the so-called "precautionary principle": "Where there are threats of serious or irreversible damage, lack of full scientific certainty shall not be used as a reason for postponing cost-effective measures to prevent environmental degradation."

The precautionary principle had long been anathema to promoters of biotech because, as they argued, it threw science out the window and opened a dangerous door to put new technologies such as theirs into a bind of "guilty unless proven innocent." No technology could be scientifically proved to be 100 percent safe, they retorted, especially new technologies that were only in their adolescence. Besides, there was a little bit of risk in everything. As one scientist active in testifying before European legislative regulatory bodies put it, "Crossing the street involves some level of risk that you'll be hit by a bus. If we stopped people from crossing the street until we could guarantee zero risk from that, people would be stranded on the sidewalks their whole life."

Wrestling with the Precautionary Principle

Proponents of GM crops and U.S. government regulators labored to undermine the precautionary principle and erode it as fear mongering rather than science. Meanwhile, the parties to the Convention on Biological Diversity—which the United States had signed but failed to ratify— set their sights on applying principle number 15 to biotechnology. The ad hoc working group on biosafety met six times between 1996 and 1999 to try to iron out guidelines for the transboundary movement of creatures "resulting from modern biotechnology that may have adverse effect on the conservation and sustainable use of biological diversity." The working group avoided the use of the more contentious term GMOs (genetically modified organisms) in favor of the acronym LMOs, for "living modified organisms."

The working group's meetings were not without disagreements, and it soon became clear that it would not be easy to draft a document that would satisfy everyone. After four initial meetings failed to make progress, an additional two meetings were scheduled and the committee was given another year to try to find common ground.

At the sixth and final meeting—held in February 1999 in Cartagena, Colombia—a compromise proposal was finally put forward. By that time, the negotiating parties had fractured into distinct groups, each representing its own set of concerns with little mutual agreement or indeed overlap.

As expected, wrangling over the precautionary principle and how far it could be applied figured prominently in the debates. The large but less powerful "like-minded group," representing primarily developing nations and China, argued for broad and robust inclusion of precautionary protections. The European Union held the middle ground, at least in the beginning, but as the stalemate continued the EU negotiators began to back down in seeming acceptance of weaker guidelines. But the hard-line "Miami group" (named for the locale of a prior meeting)—made up of the curious combination of the United States, Canada, Australia, Argentina, Chile, and Uruguay, brought together by the common denominator of being significant GM seed and crop exporting countries—interjected a new issue into the fray.

That issue was trade, and the Miami group was not willing to budge on trade regulations. The precautionary principle, they argued, and the entire biosafety protocol itself, threatened to impinge on free trade rules guaranteed by the World Trade Organization (WTO). They loudly proclaimed that, without explicit language spelling out the hierarchy of WTO rules preempting the biosafety protocol, there would be no movement forward. While they were at it, the Miami group also pushed to limit the biosafety protocol to only those LMOs intended for deliberate release into the environment such as seeds, excluding food and feed as well as processed products derived from LMOs.

At 3:30 A.M. on February 24, after four long days of negotiation, it became clear that no compromise could be reached and that consensus would be impossible. By blocking consensus, the Miami group had won the final round, it appeared, as delegates packed their bags and went home. As a nonparty to the Convention on Biological Diversity, the United States had been allowed to participate in negotiations but had no voting rights. Despite being the biggest nonparty to the convention, it seemed like U.S. trade interests had managed to ruin the party for everyone else.

The Tide Turns and a Biosafety Protocol Emerges

But some victories start out small, and the voices of a number of the "like-minded" developing countries made themselves heard at Cartagena despite the apparent defeat. In retrospect, Tewolde Egziabher, the director of Ethiopia's environmental protection agency, emphasized that "it was important that ultimately we did not succumb to the pressure of the Miami group." (The following year, for his work in safeguarding biodiversity, Egziabher was the recipient of the international Right Livelihood Award, commonly known as the "alternative Nobel Prize.") Representatives from Zambia complained that biosafety was supposed to be the goal of the protocol, not trade. Kenya's negotiators explicitly stated their intention that in any future talks, "international interests and global welfare will take priority over short-term interests and short-term profits." Cuba directly chastised the Miami group and ominously warned that "history will not forgive us." NGOs around the world reacted with predictable fury, and although global agreement on the regulation of GMOs seemed to have been laid in its grave, public attention to the issue of GMOs failed to die down.

In the final hours of the meeting, the Miami group, using voting member Canada as their mouthpiece, had proposed an eighteen-month moratorium on the talks. This would allow time enough to let things cool down (and for an overdue and overly restrictive revision of the WTO Agreement on Agriculture that would favor the Miami group's interests to take place, argued critics). A last-ditch meeting was set for January 2000 in Montreal, Canada. However, after the dispiriting failure at Cartagena, few expected that there would be any lifeblood left in finding common ground on the global regulation of LMOs.

But when the group convened in Montreal, something had shifted. Public concern over GM food was on the rise, as evidenced by protests by people wearing tassel-hatted GM corncob costumes and civil disobedience by people dressed as butterflies opposing Bt (who had made their debut at the widely publicized Seattle WTO protests at the end of 1999).

In this new light, the biotechnology industry saw a silver lining in passage of the biosafety protocol: Global consensus, or at least any appearance of it, would reassure consumers that GM food was safe and well looked after. Further bullying at the negotiations would only bring added attention and public suspicion and would preclude the kinder, gentler image of third-world salvation through GM food that the industry was brewing behind the scenes. Besides, the EU had already backed down from its most contested issues at Cartagena, and it was clear that there would be no more second—or by this time, sixth or seventh—chances to pass the protocol.

As dawn was approaching in Montreal on January 29, 2000, a deal was struck that everyone was willing to agree on. Food and feed were included in the protocol's restrictions, albeit in a lesser form than some desired. Processed food products, such as peanut butter and oils, were not. The precautionary principle came through the protocol loud and clear, from preamble to practice. And, in a nod to the stubborn and powerful Miami group, it was clearly spelled out that the WTO rules would prevail over the protocol: "This Protocol shall not be interpreted as implying a change in the rights and obligations of a Party under any existing international agreement." Or perhaps the spelling out was not so clear after all; the following paragraph seemed to take an about-face, stating: "The above recital is not intended to subordinate this Protocol to other international agreements." The issue of whether the WTO could force countries to accept GM foods over their own policies and protests was still a few years off.

Supermarkets Jump on the Anti-GM Bandwagon

Politicking and protocols aside, it was really the voices of the masses that mattered, and consumers in Europe defiantly did not want GM food on their dinner plates or even on the shelves of their supermarkets. In 1998, the small British supermarket chain Iceland had been the first to remove all foods containing GM products from its private label brand. It did not take long for most of the major food retailers across Europe to follow suit. A consortium between seven of Europe's biggest supermarket chains—Sainsbury's and Marks & Spencer in England, Carrefour in France, Delhaize in Belgium, Migros in Switzerland, Effelunga in Italy, and Superquinn in Ireland—quickly sprang up to collectively source non-GMO ingredients for their private label brands. A year later, nearly all supermarket shelves across Europe had been wiped clean of private label products containing GMOs. Supermarket executives professed that they were merely reacting to consumer pressure, but later, industry proponents accused them of promoting their own economic interests at the expense of misleading consumers into an anti-GM frenzy in the process.

The industry argument went like this: The European grocery market had been stagnating prior to the GM bans. This position was reinforced by an article published in the journal *Nature Biotechnology* by Nicholas Kalaitzandonakes, the director of the Economics and Management of Agrobiotechnology Center at the University of Missouri (the home state to Monsanto's corporate headquarters), and Jos Bijman, a researcher at the Agricultural Economics Research Institute in the Netherlands. This stagnation in turn had led to market consolidation, the two writers argued, whereby the bigger grocery fish had gobbled up the smaller ones until only a few large supermarket brands grew to dominate the food chain. With growth through consolidation reaching its limits, the dominant multinational grocery chains used the development of private label lines—basically, exclusive "store brands"—as a way to encourage further growth and increase their brand loyalty at the same time. It was in these private label store brands that the large chains introduced their exclusively "GM-free" fare.

But if they were not driven by consumer demand, why would these big players take the burdensome and potentially costly step of removing GM ingredients from their products, when the competing brands of foods did not? According to Kalaitzandonakes and Bijman, "With little manufacturing capacity of their own and sufficient buying power, retailers could pass the bulk of such added risks and costs to their ingredient suppliers and manufacturers." Faced with on-shelf competition from non-GM products, those same food suppliers and manufacturers would have to comply in their own brands, and at their own greater cost, which would then be passed on to the consumer, potentially giving the private label brands an economic advantage. The large supermarket chains seemed to be beating the food manufacturers at their own game.

In their paper, Kalaitzandonakes and Bijman suggested that perhaps it was not consumers themselves who had initiated the European distaste for GM foods in the first place. Instead, greedy supermarket retailers had strategically exploited consumers' nascent fears, and by removing GM ingredients from entire brand lines, had forced them to choose only non-GM items. Such a claim of exploitation was, of course, based on the notion that given a side-by-side comparison, there were European consumers who would actively choose the GM product over the non-GM version.

It did not matter that Kalaitzandonakes was a proponent of unrestricted GM trade and a critic of the Cartagena Protocol. (He would go on to write an article entitled "Cartagena Protocol: A New Trade Barrier?" in the magazine *Regulation*, published by the conservative Cato Institute think tank.) It did not matter that, when Swiss consumers were given a choice between GM and non-GM versions of the same product, they staged vociferous protests when bags of dog food containing GM ingredients appeared on the shelves. Within a matter of hours, retailers had swept their shelves clean of the offending item, learning from their mistake that even GM products intended for pets could incite disruptive protests. (Switzerland was free from EU constraints and its own GM labeling policy was already in effect.)

Cognitive Misers and Scientific Decision Making

The issue of choice aside, did European consumers really know what it was they were rejecting? In an ideal world (or so the theory goes), consumers would make decisions about whether to accept or reject a new technology based on scientific literacy. They would gather facts and data about the new development and its applications and use such knowledge to come to an informed and rational decision. But in reality, this approach applies only to a small and dedicated subpopulation.

Most of the rest of us arrive at decisions about new technologies via what is known as the "cognitive miser" model, in which conclusions are drawn from only the most basic and minimal information, collected from the most convenient and readily available sources. Those who subscribe to scientific literacy as the only sound route to decision making bemoan a lack of scientific understanding and a knowledge deficit in the general population. (For example, pro-biotech scientists are fond of citing the example of a European survey in which a majority of respondents indicated a belief that only GM tomatoes contained genes, while conventional tomatoes did not.) However, the rest of the populace who are cognitive misers know better than to invest too much time or effort in seeking information upon which to base their decisions.

According to social scientists who study collective decision making, for cognitive misers, investing as little time and energy as possible in gathering information about something as abstract as genetic manipulation of food crops isn't just a cop-out: It actually makes logical sense. Given the slew of "cognitive shortcuts" the world now offers—from religious doctrines to media infiltration to the growing number of NGOs and industry groups given free range on the Internet—is it any wonder that most consumers take these shorter routes when it comes to making decisions about GM foods? The downside, of course, is that consumers sometimes end up making decisions with too little information.

The cognitive miser model can in some ways account for the vast differences in consumer acceptance of GM food observed on different sides of the Atlantic. In Europe, exhaustive and repeated studies of media coverage of biotechnology have indicated that the salience, or prominence, of biotechnology in the mass media across Europe dramatically gained ground beginning in the 1990s. By the mid-1990s, "green" biotechnology— associated with the genetic manipulation of plants and foodstuffs—was primarily framed by the European media as problematic and risky, in contrast to more health-related "red" biotechnologies, which the media promoted as promising progress and beneficence. Nonetheless, overall media coverage in Europe did not warrant accusations of scare-mongering and in some cases indicated more positive coverage than in the U.S. press. In contrast, however, no such spike in salience or shift in framing was observed in the American media, though biotechnology in general was not entirely absent from coverage there.

Some might argue that the media merely represents rather than actively shapes debates taking place in the public realm regarding a new technology such as GM food. Certainly, national and cultural values are also at play in the cognitive miser model. For Europeans, the topic of genetically modified food brought to the fore a latent historical tension between science and technology on the one hand, and nature and the environment on the other, that had been brewing in the European subconscious since the days of the scientific revolution.

Historical Tensions and the European Subconscious

Take Switzerland again as a prominent pan-European example. This small nation of highly educated, scientifically inclined, deeply democratic, multilingual, and cross-cultural individuals harbors both the largest per capita number of Nobel laureates and the highest per capita consumption of organic food in Europe. Greenpeace stages some of its largest and best-organized anti-GMO campaigns on the lawns of multinational biotech headquarters such as Novartis and Roche. In 2004, 44 percent of the Swiss labor market worked in areas related to science and technology or occupied executive functions in these fields.

Yet many of these same technically oriented power-broker folks march through the streets of Zurich each year in ceremonies such as the rite of Sechseläuten, one of the city's most colorful celebrations, in which they dance around the fictitious Böögg, a snowman who is set on fire in an attempt to bring on an early spring. Throughout Switzerland and much of the European continent, seasonal rituals such as these are alive and well, embodying connections to an earlier prescientific way of life that integrates and in fact requires a view of nature as driven by a will of its own. Standing side-by-side with respect for precision technologies and high degrees of scientific literacy, these rituals honor a connection to nature's cycles and spirits far beyond science or technology's grasp.

Thus, tension exists between embracing science and technology as tools toward progress and respecting an organic worldview that elevates the power of nature over man. The roots of this tension can be traced back to the scientific revolution that swept across Europe beginning in the sixteenth century and lives on today in the continent's ambivalence toward GM food. Fundamental to the scientific revolution was the replacement of an earlier, organic worldview, which saw nature as a spirited, nurturing force with a will of its own, by a mechanistic philosophy that posited nature as an inert repository of resources to be put on a rack, constrained, and molded into new and artificial forms by the hand of man. Accompanying the rise of modern science was a new scientific methodology. No longer was nature to be approached as a benevolent collaborator; instead, all preconceptions and emotional underpinnings were to be purged in the objective and dispassionate investigation of the natural world. Nature became natural resource, matter was debased of spirit and sovereignty, and the living world was transformed into raw material to be probed and prodded by the reason of man.

This historical tension was dormant for centuries, but the issue of GM food in Europe brought it to a schizophrenic crisis. After all, here was a new and profitably promising technology, stemming from some of the most influential science of the Western world, but that potentially threatened to disrupt the most sacred of organic covenants—that linking soil, seed, and the intimate relationship between plants and people that had sustained human civilization since its inception. The Cartagena Protocol could be seen by some as an attempt to quell this crisis, to bring some rationality and order to the topic without dismissing the sacrosanctity of respect for nature. But by the time the Cartagena Protocol made it off the table in Montreal, the issue of GM food in Europe had escalated into an emotional frenzy for both its opponents and its supporters.

Asked why Europeans react so vociferously to the mention of GM food, until this point in time a relative nonissue for American eaters, a biochemist active in the GM food debate in both industry and academia pins it on one word: emotion. "Not that emotions are a bad thing," she concedes, "but they just don't have a place when it comes to science." To demonstrate this schism that raises emotion over reason, she cites consumer rejection of vitamin B2 supplements produced in bacteria using recombinant DNA technologies, even though genetic modification produces a purer and biochemically identical product to that manufactured synthetically.

"Thinking from the belly" is how a scientist active in European policy debates puts it, with just a tad of irony. He too agrees that emotion is a necessary human element, but he feels that in the debate over GM food, emotion has "eliminated the brain, which is very dangerous," especially when it comes to tinkering with plants. "You can do anything you want with mice, which is crazy, but if you touch plants you are in trouble, because plants in the Germanic mind are Mother Nature herself," he explains.

Other scientists are even more succinct in their denigration of emotion and its ties to a prescientific, romantic worldview. A prominent European neuroscientist believes that the opposition to genetic engineering has been "turned into a sect like a religious movement based on romanticism, where the ugly is allowed." He asks, "Is there one scientific paper proving that organic farming is healthier or more ecological? The whole thing has become based on belief," he continues, in contrast to reason. "Make me a graph of where the opposition to GM food originates on the map of Europe; you'll see that it looks just the same as the map of witch burnings in the time of cornfields and ghosts. It's all in the unconscious," he explains, referencing imagery related to a Lenten fertility ritual popular across much of Belgium, Germany, Austria, France, and Switzerland.

This tension between modern science and its translation into industrial agriculture and biotech foods on the one hand, and a European history that forges an emotional connection between people and nature through food and farming on the other, crops up readily in discussions with those who organize the opposition to GM foods in Europe. Many of those who work fervently to oppose the introduction of genetic engineering into the environment and the food supply reflect such views. They express strikingly similar sentiments to Victor Frankenstein's early connection to nature described in Mary Shelley's eighteenth-century novel *Frankenstein*. In the metaphor of "Frankenfoods," anti-GM activists in Switzerland see parallels between Frankenstein's earlier romantic relationship to nature and his eventual fall from grace at the hands of modern science—a fall in which emotional detachment and excessive zeal for manipulating material nature led to the disastrous creation of a monster that roamed out of control, destroying all it touched. It is this loss of emotional reasoning and disrespect for the autonomy and sacrosanctity of nature that anti-GM activists in Europe warn against and angrily pin upon their scientific adversaries. And it is this aspect of the Frankenfood metaphor, more so even than the fear of the monster itself, that resonates with so many opponents of GMOs and drives them to action.

Reason and Emotion in the European GM Food Fight

The Greenpeace office in Zurich, like everything else in Switzerland, is neatly organized and tightly run. A campaigner for Greenpeace's anti-GMO project walks briskly across the brightly colored linoleum floor and into a spacious office area where desks and bookcases made from sustainably harvested timber store the organization's archives. The lights are kept dim to conserve energy and to keep the office cool. It is 2003, the summer of legendary, lethal heat waves blistering across Europe, in seeming confirmation for the urgency with which Greenpeace makes its environmental pleas. Photos of recent demonstrations against GM products are tacked to bulletin boards, and a grocery cart in the corner holds samples of products found through Greenpeace's stealth testing campaigns to be unwittingly contaminated with genetically modified ingredients. The man walking through the office is proud of the work he does. With his orderly demeanor and bespectacled appearance, he seems unlikely to be pegged as a radical activist.

But his passion to prevent GM foods from becoming staples on Switzerland's grocery store shelves runs deep. Like Victor Frankenstein, he was drawn to his work through a love of and connection to nature. But where Frankenstein eschewed that connection and turned to modern science as a route to knowledge and power, this activist has honed his emotional connection to nature to a fine and focused point. His original mission in working for Greenpeace was not the lure of the organization itself, but his passion for preventing what he sees as the potential destruction of nature at the hands of Frankenstein-like scientists. On his regular commute from Zurich to Bern, the seat of Switzerland's national government, he would stare with awe at the expanse of yellow rapeseed fields each spring.

"When I saw those yellow rapeseed fields in spring, I would want to cry, because I felt that in some years, even this will be totally artificial plants, constructs made by humans rather than nature, and where are we going with that? For me, that is the end of nature." And that is how he came to be a campaigner for Greenpeace. "Up until now, we have been taking more and more the place of nature, with our streets, our buildings, our fields, and now we are even changing the DNA of living beings," he said. "This is something that we really should not do, because this is the end of nature. We should not play God, not only based on ethical issues but also based on the ethics of environmental protection. It makes sense that a rose cannot make babies together with a horse!"

The Greenpeace campaigner and his coconspirators—if one can call such a seemingly rational, earnest young man by that appellation—have organized regular "manifestations" (as the German term for street demonstrations translates) to help bring public awareness to their quest to ban GM field testing in Switzerland. In one such event, a large number of female activists identically dressed as schoolgirls were assembled in neat lines on the lawn of Novartis's corporate headquarters. Each held a naked baby doll with a funnel inserted into its mouth, through which kernels of corn were fed.

Despite their vivid antics, members of non-governmental, grass-roots organizations such as Greenpeace are not the only ones playing the emotion card. Scientists routinely criticize their opponents (the "greenies") for drawing on the emotions of a naive public. Yet when asked about the controversy over GMOs in their country, many of these same scientists fly into an emotional fervor that often makes NGO activists seem tame in comparison.

An esteemed molecular biologist who left academia to found a private European biotechnology venture is reluctant to speak publicly about agricultural biotechnology. He feels that the science has been misunderstood and is often wrongly interpreted. But when he does speak, he has sharp, passionate words for his opponents. He feels that even members of reputable environmental groups (to whom he admits having given financial support in the past) have acted in bad faith with regard to biotechnology. He thinks that these environmentalists have pursued an agenda that he says is "simply trying to fool the public by increasing the idea of risk, by increasing the notion of fear, by deliberately trying to introduce fear in a public that was so ignorant that they were victims." Although he admits that these "greenies" (as he refers to them) are "not some kind of lunatic ayatollahs," in his mind, they must either be "ignorant, or of bad faith" to promote such an extreme agenda.

American Apathy Toward GM Food

In Europe, the emotional battle lines between the supporters and critics of biotech foods have been drawn deep. In the United States, on the other hand, there is no messy history regarding science and technology that compares to the tensions found in Europe. In the United States, there had been a backlash against nuclear power stations in the 1970s, and even the Internet had found a few detractors in its nascent years. But with the exception of the lone voice of writer and activist Jeremy Rifkin, to Americans, technology and the "big science" it was tethered to meant progress, growth, and world dominance. Big science had put Americans on the moon, and more recently the Human Genome Project had demonstrated the promise of the race for new technologies that pushed forward the frontiers of the life sciences. Genetically engineering agriculture was just more of the same, using human genius to improve upon the bounteous raw materials found in the natural landscape.

Americans also had a decidedly different relationship to food than Europeans. Europe had its foie gras, Wiener schnitzel, and antipasto. In contrast, Americans drew their culinary inspiration from baseball, hot dogs, and apple pie, as a famous advertising campaign for Chevrolet once sang out. While Europe was the birthplace of the Slow Food Movement—founded to preserve traditional food customs and keep food at the center of community—the United States was addicted to fast food, which by the dawn of the twenty-first century was getting faster every day. An American meal that was allotted one hour of prep and cleanup time in 1965 garnered only thirty-one minutes by 1995. By 2007, 90 percent of Americans ate food each day prepared away from home, and drive-throughs were booming. Consumers shelled out $110 billion on fast food in 2000, compared to only $6 billion in 1970.

Not only was food in the United States getting faster; it was also getting cheaper, another food value that divided Americans from Europeans. By the end of the twentieth century, Europeans spent between 15 and 20 percent of their household income on food, while Americans spent roughly half that amount. A steak in Denmark cost more than twice as much as the same side of beef purchased in the United States, while a loaf of bread baked in France garnered a 25 percent premium compared to those from American hearths. Lush agricultural subsidies and new technologies that turned excess commodity crops such as soy and corn into fast-food ingredients, like trans fats and high fructose corn syrup, kept Americans hooked on processed food. By the end of the 1990s, Americans were addicted to cheap corn and soy, although they often never knew they were eating it because its form was so processed in so many products.

Not surprisingly, these were the very crops that biotech giants like Monsanto and Syngenta had set their sights on a few decades earlier. And so, given the convergence of these factors, it was no surprise that Americans would gobble down GM food without batting an eye, notwithstanding the fact that the GM foods were delivered to them unlabeled and disguised within a healthy dose of their favorite calorie-laden processed foods. As one young European anti-GM activist put it, "If Americans are willing to eat McDonald's special sauce, it's no wonder they don't care if it's genetically modified or not."

Nor should it have been surprising that these so-called Frankenfoods were so vehemently rejected by European consumers. To Europeans, the Frankenstein analogy was a reminder of the historical violence of the betrayal of nature by science, a moral and cultural offense. To Americans, raised in a far more violent cultural context, the threat of Frankenstein paled in comparison to the Terminators and Pulp Fictions that Hollywood bombarded them with. And besides, in the United States, scientific hubris was a good thing, not just acceptable but necessary as a driving force behind the nation's global dominance in world economics and trade.

Which was precisely why U.S. officials were adamant that the romantic European concerns for nature and the environment embodied in the Cartagena Protocol not interfere in any way with the global trade of one of America's most precious commodities, GM food. After the adoption of Cartagena by more than 130 countries in January 2000, the United States was poised to use its might to iron out any uncertainties and prove that its provisions would bow to the WTO when necessary. And thus began a game of cat-and-mouse gastronomy, with the issue of European acceptance of GM food imports serving as a political Ping-Pong ball tossed to and fro across the Atlantic.

The StarLink Saga

The ink had barely had time to dry on the Cartagena Protocol when the United States threatened to press its cause at the WTO. But fate intervened, and in the summer of 2000, the first major GM food scandal hit the United States. StarLink was a genetically modified strain of Bt corn that had been developed by Aventis Crop Sciences (later acquired by Bayer) and had received approval for commercial planting from the U.S. Department of Agriculture (USDA). However, its questionable performance on tests to determine its potential to cause allergic reactions prevented it from being approved for human consumption. Thus, farmers were allowed to plant StarLink under the provision that it be used only for animal feed or other nonhuman uses. In retrospect, this was a recipe for failure, given the scale of corn transport and processing in the industrialized agricultural channels of the United States. But at the time, Aventis assured the USDA that a system was in place that would allow total segregation of StarLink from the food supply.

Such an unrealistic scenario was like manna from heaven to anti-GM watchdog groups such as Friends of the Earth. The double standards associated with StarLink provided them with the perfect opportunity to prove that containment was impossible and that renegade GM foods would lead to widespread contamination, with the potential for allergies to boot. In September 2000, a random check of Kraft's Taco Bell brand of taco shells turned up a positive test for the presence of the unapproved StarLink corn. Not long after, StarLink contamination was detected in taco shells produced by Texas-based Mission Foods, which quickly pulled all of its products containing any corn at all from supermarket shelves. Complicating the matter, Mission Foods produced store-brand taco shells for many of the major supermarket chains, triggering a domino-effect recall of corn products under the Safeway, Food Lion, and Shaw's labels as well.

Attempting to avert a major public relations catastrophe, Aventis quickly bought up the remaining StarLink harvests, but several million bushels still went unaccounted for. In the meantime, two dozen consumers had come forward, insisting that they had experienced allergic reactions after eating the contaminated taco shells. While blood samples taken from several of the purported victims did show signs of allergic reactions, an investigation by the Centers for Disease Control concluded that the allergy had not been to the StarLink protein, and the case was terminated with the official confirmation that no allergic reactions had been caused by the contraband corn. With the case officially closed and StarLink off the market, the USDA vowed never to issue such a split approval again, and the public went back to eating taco shells. Concern over GM food on the part of American consumers was all but forgotten.

Europe Stalls, the WTO Strong-Arms

By 2001, the specter of a showdown at the WTO was looming again, as European Union member states continued to reject attempts by the EU

Commission to restart approvals for GM foods. In mid-2001, EU Commissioners assured the United States that new traceability and labeling laws would be put into effect and that the ban would be lifted within a matter of weeks. Weeks turned into more than a year, and it was not until October 2002 that the new EU regulations finally came into effect, in theory allowing the resubmission of GM applications that had been lying dormant and gathering dust for the past four years.

U.S. trade representatives were still skeptical of the EU Commission's ability to follow through, and they spouted statistics of the $300 million annual price tag that the European ban was costing them. On May 13, 2003, the trade representatives joined their counterparts from Argentina and Canada—two of the largest GM-growing nations besides the United States—in making a formal request to the WTO Dispute Settlement Body for consultations with the EU on the issue of GM foods. Their timing was impeccable. While it may have seemed on the surface that they had waited patiently long enough, the date preceded by exactly one month the ratification of the Cartagena Protocol by the fiftieth country, the tiny island nation of Palau, triggering a ninety-day countdown until its provisions went into force.

Over the summer of 2003, the U.S. administration ramped up its message, inserting references to hunger and famine striking Africa as a wedge between what President George W. Bush called the "artificial obstacles" put up by Europe and the humanitarian good that GM food could bring to starving countries. Anti-GM European policies, Bush contended in a speech before the annual biotechnology industry conference in June 2003, were intimidating African nations, bullying them out of investing in GM technology, and heightening the hunger already plaguing their citizens. "For the sake of a continent threatened by famine, I urge the European governments to end their opposition to biotechnology," Bush intoned in closing.

Accusations of bullying aside, the Bush administration was not going to rely on Europeans' humanitarian leanings to solve the GM impasse. At the end of August 2003—just two weeks before Cartagena went into effect—

the WTO established a panel to investigate the complaints filed by the United States, Canada, and Argentina against the European GM policy. While it had been rumored that Egypt would join the GM powerhouses in the suit, the Egyptians pulled out at the last minute, swayed by domestic opposition within the country. In a quick bout of retaliation, the United States immediately suspended free trade talks with Egypt, shocking a nation that had been courted as a friend and party to the lawsuit in the name of developing-nation needs. This sharp smack in the face surely sent a message to other less powerful nations about the rules for playing with big brother. "If you're going to do a free trade agreement with someone, it's important that the people you're talking to are going to be able to deliver. They told us one thing and did another," chastised a U.S. trade official in explaining the decision to mete out punishment.

Over the next two years, the WTO investigation snaked its way through an ongoing series of bureaucratic nooks and crannies, gathering *amicus curiae* ("friend of the court") submissions from numerous countries and councils, and pestered by petitions signed by more than 100,000 citizens in ninety countries and 544 organizations representing 48 million people opposed to eating GM food. In 2004, the European Union methodically approved its first GM imports since 1998, and then its second. But the WTO case continued. At the end of 2004, the EU shot back at the United States and Canada, filing its own dispute against them in the same arena, challenging sanctions imposed in retaliation against the EU's ban on hormone-treated beef. The food fray was on, in full force.

After a scientific advisory board was called in for consultation, by the end of 2005 it was rumored that the WTO decision was imminent. Rumors also held that the United States would be declared the victor, although with the approval of GM crops already under way once again in Europe, it might be seen as a hollow victory. But a victory it was, nonetheless. When the final report was finally issued in September 2006, it was the longest ruling in the history of the WTO, and within its 1,087 pages it clearly declared the United

States a winner on all counts. Exhausted and determined that many (but not all) of the WTO declarations were moot points anyway, the EU took the unusual step of deciding not to appeal, and the WTO case concerning GM foods came to a close.

Europe and the Future of GM Food

Although the legalities may have been decided by the WTO, the jury is still out on whether European consumers will eventually accept GM foods. Will Europe's stonewalling on GMOs, like the opposition to recombinant DNA in the 1970s, become relegated to the annals of history, another knee-jerk reaction to a new science beleaguered by Luddite fears? Or will Europeans keep up the food fight, refusing to refine their palates to conform to international trade treaties?

The experience in Switzerland, which is not a member of the European Union and thus not party to the suit, may be indicative of the direction that the EU insistence on labeling may take. Back in 1995, Switzerland was one of the first countries in the world to institute a labeling law. Yet more than a decade later, GMOs remain conspicuously absent from supermarket shelves and consumers' dinner plates. If anything, consumer sentiment in Switzerland has only grown away from the high-tech interventions and toward the "back to nature" approach of organic farming symbolized by a yellow and green "Bio" label calling out organically grown foods on supermarket shelves. Will the rest of Europe follow suit, digging in their heels even more vigorously once confronted with the daily and ubiquitous presence of GMOs on supermarket shelves? Will Europeans remain the isolated outsiders taking an increasingly minority view of GM foods?

It seems unlikely. Although Swiss consumers rally against corporate globalization, particularly when it comes to food, there is no denying that the earth's economy and indeed the world food supply is increasingly knitted together within a tight meshwork of economic, environmental, and social forces. Within this changing landscape, it seems unlikely that Switzerland or Europe as a whole can hold out a singular position on GMOs for very long, without greater world forces weighing in. The mountains and valleys of Switzerland may have preserved the unique and bucolic nature of Swiss identity, environment, and farming for hundreds of years, but globalization has become a force stronger than the bedrock of the Matterhorn—indeed, stronger than nature itself.

The fate of GM food in Switzerland and the European Union will not proceed in isolation. Although the debate may have originated as a United States/Europe conflict, the tentacles of the trade ruling extend deep into the political psyche of smaller, less powerful nations that may have more at stake with regard to GM foods. It is here, in the stomachs and on the shelves of the hungry majority of the world, that the real food fray remains to be fought.

CHAPTER 4

A Kinder, Gentler GM:

Will Biotech Seeds Save the World?

Anti-biotechnology activists in Europe had played the emotion card, and it had worked. With images of Frankenstein hovering over GM food, Europeans weren't just scared: They were angry. But now, on the other side of the Atlantic, as Bush administration officials gathered documents to pursue the World Trade Organization (WTO) case, the biotech industry was determined to do the activists one better. Rather than wowing the public with science—an approach that had backfired in the ice-minus fiasco (discussed in Chapter 1) a decade earlier—the biotech lobby would tug at the public's heartstrings. If Europeans wouldn't accept GM food, and unaware Americans merely gulped it down passively, it would be left to the third world to embrace it. In the eyes and minds of the biotechnology industry, citizens of developing nations would make wonderful poster children, as the hungry and willing recipients of the humanitarian beneficence that biotech promised.

Biotech's Humanitarian Facade

In the spring of 2001, the Council for Biotechnology Information, a group comprised of "the leading biotechnology companies and trade associations," introduced a television advertisement as part of its new "Why Biotech?" campaign. In the ad, a voice-over extols the virtues of biotech food as images of half-clothed African children roll by. Then Maria Brunn, a young American cancer survivor, skips across the screen on her way to sports practice. "Thanks to biotech medicines, Maria's cancer is in remission and she's back on the team," intones her beaming, blonde mother. Images of scientists clad in white coats and staring through microscopes flash across the screen.

As the ad winds down from its focus on first-world medical applications of biotechnology, the camera pans across a field of stooped Asian peasants, faces shaded by their pointed bamboo hats. "From medicine to agriculture, biotechnology is providing solutions that are improving lives today, and could improve our world tomorrow," an authoritative female voice declares. After a breathy pause upon the word "today," the camera zooms in on a female farm worker, her distant stare obscured only slightly by the child slung on her side. As the word "tomorrow" is articulated, the mother's gaze lifts, foreshadowing a future filled with biotech food crusades that will take place in far-off fields.

Poor Farmers, Speaking for Themselves?

This advertisement was the timid beginning of a campaign in which generic (and often purportedly fictitious) farmers from the developing world were paraded in front of well-fed Westerners in the name of GM food. At the same time they were slipping themes like hunger and poverty into the mainstream marketing of biotechnology, industry insiders were also swaying their message to include connections to humanitarian missions. The theme of humanitarian aid via GM food often reared its head, at venues ranging from United Nations Development meetings to the raging, party-like atmosphere of cocktail hour hospitality suites at the biotechnology industry's annual conference.

Suddenly, it seemed that every other industry scientist or representative had met a woman named Mei, Ling or some permutation of an Asian-sounding name. Burdened by hard labor in the fields of her native land, she struggled to feed her children. With a wave of the magic wand of biotechnology, these industry insiders enthused, this woman's problems could be gone: her crops flourishing and pest-free, her income enhanced so that she could send her children to school, and her hands freed from the endless hours of labor imposed by the small and burdensome scale of rural agriculture.

Of course, almost none of these scientists had really met the Meis and Lings of their marketing campaigns. In many cases, these images—like the Asian woman in the Council for Biotechnology Information advertisement, working with one hand in the fields while toting a child along in the other— were merely icons, references to a kinder and gentler approach to GM food consistent with corporate marketing trends of the time that reached beyond the bounds of biotechnology. Indeed, the exact same image of the woman in the field that ended the Council for Biotechnology Information ad reappeared again that year in another television advertisement. This time, it was for the airplane manufacturer Boeing as it launched its "Forever New Frontiers" campaign in 2001 by begging the question, "What will tomorrow bring?"

Soon, darker faces were added to the repertoire, and names and voices were added to allow those whose future had been transformed by biotechnology to speak for themselves. Monsanto's website was spiced up with short video testimonials from farmers as far away as Burkina Faso, South Africa, and the Philippines.

With the advent of online video sharing websites, Monsanto's message spread even further afield. With all references to the company removed, these same videos were posted on the Internet video site YouTube, under the suspicious screen name "Sidewinder 77." Here—amid homemade videos ranging from "Fast Cars" to "How to Hit a Home Run"—Monsanto's clips sprouted up as grass-roots, homemade pleas from third world farmers on behalf of the good of GM.

"My name is Thandiwe Myeni, I've got five kids, my husband is no more," begins one solemn testimonial. "We had a huge problem—the problem of the insects, the worms . . . and in 1999–2000 I wanted to try this Bollgard to see . . ." she continues, referring to Monsanto's proprietary variety of Bt cotton. "The Bollgard is a little bit expensive than conventional. . . ." Her voice trails off, and suddenly, the camera cuts to a view of the same woman sitting at a desk, perhaps writing up her profit statements or receiving a check. "But, at the end of the day, we save, you save a lot of money . . . and for my kids, I use the money from cotton to take them to universities."

Such an earnest and concerted approach was sure to stop any opposition in its tracks, taming even the scariest Frankenfood monster into a friendly, benevolent force curing humanity's ills. But were GM crops really a solution for poor farmers? Could biotechnology be the magic bullet that finally put world hunger to rest?

Norman Borlaug, Voice of the Green Revolution

Just in case the voices of poor farmers themselves weren't enough to convince a wary public that GM was good for the world, Monsanto added to its online repertoire a series of authoritative lectures from "experts who research and study the benefits of biotechnology in agriculture." Among this list was a brief endorsement of "the benefits and safety of biotechnology" from Norman Borlaug, identified as Nobel laureate, USA.

Born in 1914 to Norwegian immigrants who had settled in Iowa, Borlaug had received the Nobel Peace Prize in 1970 for his work pioneering the legendary Green Revolution, an international effort to increase food production in developing countries through the provision of high-yielding hybrid seeds and massive amounts of fertilizer inputs. Yet Borlaug is perhaps the quintessential "forgotten benefactor of humanity," as a 1997 article in the magazine *Atlantic Monthly* labeled him. Few today can clearly recount the story of an agricultural revolution in the 1960s and 1970s that reshaped the state of third world food systems for better, or—in the minds and lives of some—for worse.

Borlaug's Green Revolution, or at least its inception, was financed by the Rockefeller family's private foundation. With impeccable timing as World War II drew to a close, they recognized that improving agricultural productivity in the developing world could serve an important humanitarian purpose. At the same time, it could help to quell a quietly smoldering contempt among the disenfranchised masses residing there, transforming them into consumers for post-wartime surpluses of nitrogen fertilizer and preventing their conversion to socialist or communist regimes that might come courting. As the war drew to a close and U.S. foreign policy turned to facilitating the expansion of capitalism and American multinational business interests, the economic development of the third world became a pressing priority. The persistent shadow of hunger and poverty in these regions acted as both a barrier and an embarrassment to forging forward with the capitalist regime. Thus, it was high time for a humanitarian intervention.

Early Green Revolution Successes in Mexico

The first target was Mexico. The country seemed an obvious starting place, with a newly elected government under President Manuel Avila Camacho, who was an advocate of industrialization with farm roots himself and a new wartime friend of the United States. In Mexico, successive epidemics of the devastating wheat disease stem rust, which led to shriveled seed heads and often brought the crop to the ground before harvest, had devastated grain yields. In 1944, Mexico imported half the wheat it needed to feed its citizens. That same year, Norman Borlaug arrived and set to work identifying and hybridizing new rust-resistant cultivars of wheat that were short, stout, and responded with incredible vigor to heavy irrigation and blasts of nitrogen fertilizer.

One of the problems with conventionally bred grain crops was that while they did increase production in response to water and fertilizer, their tall stems and abundant seed heads tended to make them topple over, spilling their bounty to the ground in a process called lodging. But Borlaug was able to create vigorous hybrids by crossing conventional, leggy varieties, which were resistant to the pernicious rust disease and adapted to the local growing conditions, with a special short-straw variety of wheat derived from a strain called Norin-10 that had originated in Japan. Borlaug's hybrids were dwarf plants that leapt into action if heavily fed and watered, but still held their heavy seed heads high on their short, bushy stalks.

In Mexico, the impact was dramatic. Government investment in industrial agriculture infrastructure such as commercial irrigation systems and the willingness of farmers to embrace the new, input-intensive form of agriculture led to massive increases in grain production. These increases, however, were possible only because government and farmers alike adhered to the strict requirements of the new system. Hybrid varieties of seed by their very definition do not "breed true." This prevents seed saving and requires the acquisition of new seeds from a supplier each season, rather than putting aside some of the current year's harvest to replant the following year. And while the increases in yield were massive when all of the proper conditions were met, skimping on any of the fertilizer needed caused the entire system to crash, leading to only minimal increases or even decreases in production.

Farmer compliance was high in Mexico, and the requisite conditions for proper irrigation and fertilization could be achieved. For the most part, the Green Revolution—though it had not yet been called that—transformed much of Mexican agriculture from small-scale, subsistence farming to industrial commodity production. By the 1950s, Mexico was self-sufficient in wheat production, and by the mid-1960s, it had become a net exporter, shipping out half a million tons of wheat per year. A model for solving the stubborn dilemma of food insecurity seemed to have been hit upon, combining conventional genetics with the miracles of chemistry: Just add water and mix.

India Invites the Green Revolution In

From Mexico, Borlaug moved his efforts to India. There, a legacy of colonialism, combined with misguided government strategies that favored industrial development over agriculture, had led to near famine conditions by the early 1960s. Droughts and unpredictable monsoon rains had further eroded agricultural productivity to the point where India had become a caricature of third world hunger. India had managed to win its independence in the midst of the historic Bengal famine of the 1940s, in which 3 million had perished. But by mid-century, the small-scale subsistence agriculture practiced by the majority population of Indian peasants had left many of them dispirited and famished. A brief period of increased food production in the early 1950s had turned out to be merely a hollow promise, and few new approaches were looming on the horizon. In the doomsday tome *Famine—1975!*, which was published in 1967, the authors even suggested jettisoning food aid to India altogether and leaving its population to starve, so dire seemed its future for food production.

Given this backdrop, when Borlaug came knocking with his new dwarf varieties of hybrid wheat in 1965, he was boldly invited in by India's agricultural minister, C. Subramaniam, and his deputy, M. S. Swaminathan. In 1966, under raised eyebrows but at Subramaniam's stubborn direction, India imported 18,000 tons of the same hybrid seeds that had turned Mexican agriculture around in its tracks. In the first harvest season, fields planted with the new seeds increased production by 78 percent. In the second season, the increase was up to 98 percent. And by 1974, India and its uneasy neighbor Pakistan—where the hybrid seeds had managed to traverse political battle lines that the two countries had not been able to work out themselves—were both self-sufficient in wheat production. India's own scientists and agricultural experts continued to further cross the dwarf varieties to native Indian strains of wheat, ensuring a broadening and continuously increasing array of high-yielding varieties.

As in Mexico, it was not only the miracle of genetics that drove the rapid and remarkable increases in production in India. Prior to the planting of the new seeds, India had already begun to lay in place a foundation that would get the Green Revolution off to a fertile start there. In the 1950s, the Indian government had initiated various measures, including land reforms, irrigation, and fertilizer production, in response to food shortages. And modern agricultural technology was no stranger to India; the earlier colonial administration had pumped resources into developing modern technological approaches to increase crop yields, but only in cash crops destined for export, such as cotton, tea, coffee, rubber, and spices.

Thus, when the unconventional hybrid seeds arrived in India, the last remaining hurdle was to convince farmers to accept them. Discouraged by repeated crop failures and dogged by hunger, farmers did not take much cajoling, especially since the government subsidized the fertilizers and pesticides needed to complete the loop. As the first harvest season drew near, an almost instant bounty spoke for itself. Compliance would not be a problem in this case.

Promises and Failures of the Green Revolution

Despite the dramatic successes observed in Mexico and parts of India, overall global performance of the Green Revolution was spotty. In fact, it was deemed an outright failure on the African continent. Nonetheless, in certain areas of the developing world—most notably Mexico, India, Pakistan, and the Philippines—food production and yield increased dramatically in the wake of the Green Revolution, easing several of these nations out of food dependency and, some would say, into twentieth-century capitalism. For while the overarching goal promised by Green Revolution planners was the alleviation of hunger, its emphasis on intensive chemical inputs, hybrid seeds, and reliance on the acquisition of labor-replacing mechanical equipment necessarily also meant the potential for new markets for multinational agribusiness.

And thus, it was really no surprise to find Borlaug's steely face on Monsanto's website a generation later, intoning the miracles promised by plant biotechnology as another Green Revolution. After all, Borlaug had begun his career at DuPont, now one of the three "GM giants," during World War II. Although he held a doctorate in plant pathology and genetics, at DuPont his scientific efforts were directed toward wartime chemical development. Borlaug's contribution to the Green Revolution sat squarely in the arena of plant breeding, both in the laboratory and in his many years spent out in the fields. But the Green Revolution did have an insidious, detrimental side to it—a side that Borlaug might not have predicted in his enthusiasm. Decades later, the environmental toll that harsh chemicals and irrigation took began to rear its head, as yields dropped and fields fell fallow. But in the early years of the Green Revolution, the great success of the new seeds outshone any predictions of doom.

One of the characteristics that led to the massive yield increases observed in the Green Revolution was the ability to utilize conventional genetics to bring together a set of traits governing disease resistance and yield with another conglomeration of traits that led to the short, stocky phenotype that allowed this greater yield to be realized rather than wasted. Borlaug was not merely a laboratory scientist by any means, and conventional corn breeding requires not just skill and knack but persistence and long days in the field. Borlaug took to this task with zeal, not only crossing a few different strains and waiting for results but gathering diverse varieties of wheat from around the world in hopes of bringing together new combinations of traits using conventional genetic means.

Although it may on the surface appear otherwise, conventional breeding is by no means a simple task. Because wheat is self-fertile and contains both male and female reproductive structures, the immature male stamen must be carefully removed with forceps at precisely the right time from each flower to prevent the production of pollen that might slip down and self-fertilize the female pistil. After a given flower is "emasculated" in this manner, it must then be covered with a glassine sleeve to prevent any further unwanted fertilization from taking place. Several days later, spikes bearing pollen from a desired parental variety are shaken into the sleeve.

But wheat pollen is delicate and impulsive. It remains viable for a very short period, usually in the range of one to three minutes, so this procedure must be accomplished quickly and with great precision. If these conditions can be met and fertilization is successful, the seeds that subsequently develop can be collected and planted out in the next generation, and plants with the desired qualities can be selected and subjected to the same process all over again. In this way, plants bearing the genetic contributions of different and desirable parental generations can eventually be constructed and selected for.

Will GM Technology Give Rise to Another Green Revolution?

Given the labor involved in conventional breeding and the hit-or-miss genetic probabilities that constrain it, the development of directed transgenic technologies that allow for the precise transfer of individual genes from one plant to another seemed like a godsend. No long hours out in the field, no glassine sleeves and finicky pollen to deal with. Transgenic technology completely skips the sexual reproduction process that gives rise to the immense variety and unique combination of traits in our food crops. Instead, it relies on the ability of plants to regenerate whole mature specimens out of embryonic tissue, so that genes isolated from one species can be inserted into the DNA of recipient plant cells growing in a flask using a "gene gun"— literally a ballistic delivery system that bombards embryonic plant cells and tissues with tiny particles of gold coated with the gene to be transferred. In this way, the recipient plant cell is physically forced to incorporate the introduced gene into its own genome.

These transformed tissues are then treated with plant hormones that stimulate their development into shoots, a very few of which will express the desired gene in the desired manner. Many others will either express the gene in an aberrant and undesirable way or never incorporate the gene at all. Seedlings expressing a properly regulated gene are further selected to weed out plants that do not fit the overall criteria, and the surviving few of these genetically modified organisms are grown up and successively bred to determine their future suitability in the field.

In contrast to the conventional breeding that sparked the Green Revolution, transgenic technology takes place largely in the laboratory, at least initially. Proponents laud its precision. Rather than haphazardly transferring whole volumes of genetic traits—in the way that families pass on a grandfather's nose or a great-aunt's widow's peak, along with a mélange of other forms—in producing genetically modified organisms, individual genes conferring specific properties can be inserted into an already known plant background and be selected for through laboratory screening.

However, this single-gene strategy may fall short when it comes to the complex agro-ecological problems that lead to food shortages and epidemics of hunger in some of the most vulnerable regions of the world. For a company bent on conferring resistance to a favorite herbicide, or transferring a protein toxin lethal to a specific strain of insect pest, the single-gene shooting strategy surely saves time and effort. But for traits needed in agriculturally challenged regions—like drought resistance and adaptation to soils that have accumulated high levels of salt because of over-irrigation—more than one gene may be required, and such genes or groups of genes may necessitate years of research to identify and isolate. Moreover, where the development of commercially controlled, patent-protected seed stocks calls for genetic uniformity, the varying ecological environments and tyrannical climate circumstances of some of the least agriculturally productive lands in the world necessitate the genetic diversity that will ensure adaptation to the varying local conditions.

This leads into another major difference between conventionally bred crops and the genetically modified transgenic seeds being promoted as potential solutions to global hunger and food insecurity. A second important characteristic of the Green Revolution was the fact that its hybrid seeds were created and curated under public control, allowing easy access to the germplasm imbedded in them by farmers far and wide. As a public project with humanitarian ends, direct profit was not part of the equation, at least when it came to producing and providing the seeds.

When the Rockefeller Foundation and its partners initially funded the science that led to the Green Revolution, they did so through the formation of a nonprofit research organization in conjunction with the Mexican government, first known as the Office of Special Studies. In 1963, it formally became the nonprofit CIMMYT, for Centro Internacional de Mejoramiento de Maíz y Trigo, better known to outsiders as the International Maize and Wheat Improvement Center. Not long after the Green Revolution had spread to India, the International Rice Research Institute (IRRI), another nonprofit agricultural research and training center, was established in the Philippines by the Ford and Rockefeller foundations in conjunction with the Philippine government. Both CIMMYT and IRRI continue to operate today as nonprofit organizations. Their mission is the alleviation of hunger and poverty and the provision of sustainable agricultural technologies for the betterment of the poor.

Purveyors of biotech seeds may position themselves, like Monsanto does, as "just one organization that is helping the world face the challenges of increasing food self-sufficiency, meeting energy demand, improving environmental health, and creating healthier foods." But when it comes to intellectual property issues, the brusque forces that have guided the development of commercial transgenic technology are out of step with the issues confronting small farmers in far-off lands. The commercial GM traits currently available, into which so much research and development have been poured, were never designed with humanitarian intentions, and as such, it takes some contortions to recast them in that light.

At the heart of intellectual property protection is the concept that inventors should be rewarded for their unique creations and should be allowed to exercise some command over how and where these are used. Where Norman Borlaug was satisfied with rewards including a Nobel Prize and a Congressional Gold Medal, shareholders in large multinational corporations heed no such altruistic call. In theory, perhaps, but the chief goal of intellectual property protection for seeds is, in its most flattering light, a payback on the research and development investment that went into their development. At worst, it amounts to commandeering the global food supply, as some anti-GM activists in both industrialized and developing nations have charged, through the exercising of intellectual property protections, patent rights, and royalty charges. Most of the transgenic traits that have been commercialized to date were developed to reap economic rewards for their creators who already held a secure toehold on industrial agribusiness. Because of their low profit margin, seeds with traits most needed by poor, resource-strained farmers have yet to make it in this market. There is one notable exception in this regard: Golden Rice.

The Scourge of Vitamin A Deficiency and the Creation of Golden Rice

In 1999, a product emerged that on the surface at least promised to fulfill the GM revolution's humanitarian yearnings, an important touch-point in the GM food debate both historically and in the current climate of world food insecurity. Ingo Potrykus, born in Germany, was a Swiss professor of plant science who might be considered the biotech incarnation of Borlaug. Potrykus and his collaborator, Peter Beyer at the University of Freiburg in

Germany, announced that they had created a new strain of rice, fortified with B-carotene (a vitamin A precursor), that held the potential to save the lives and sight of more than a million children worldwide. Globally, vitamin A deficiency strikes with literally blinding severity, causing first night blindness as the cornea becomes dry and dehydrated, and then total and irreversible loss of sight as inflammation and infection ravage the eye from its interior. Beyond its effects on vision, vitamin A deficiency also adds to the severity of childhood diseases like measles and diarrhea, as well as depressing iron levels and disrupting the functioning of the immune system.

According to the World Health Organization, somewhere between 140 million and 250 million preschool children in 118 countries around the globe are afflicted by some degree of vitamin A deficiency. Within a given calendar year, half a million of them will go blind, and half of these again will be dead before that year is out. Vitamin A deficiency is considered a public health problem in more than half of all countries worldwide, particularly in Africa and Southeast Asia.

In many developing countries, vitamin A deficiency is largely the result of overreliance on white rice as a sole dietary staple. Lack of access to fresh fruit and vegetables and the green, leafy crops that provide requisite doses of B-carotene necessary for vitamin A production in the body are a result of poverty, lack of access to land for growing, and, some say, the Green Revolution itself.

Where the new varieties of wheat and rice introduced by the Green Revolution were incredibly efficient at providing much needed calories through their increased yield, the industrial model of farming that they rode in on displaced more diverse agricultural traditions that existed in these locales. In turn, dietary variety was reduced, as hungry stomachs stopped rumbling when the requisite number of calories came in, regardless of whether vitamin and mineral requirements were met. Reliance on the use of herbicides abolished many of the "weed" species that infiltrated growing fields and which many poor and subsistence farmers incorporated into their diet as vitamin A–rich greens. Pesticides contaminating the water in rice paddies killed off fish, ending an integrated model of rice-fish farming that had been practiced in some regions for generations and had supplied a more balanced diet.

In India, consumption of vegetables decreased by 12 percent in the decades following the Green Revolution. In Thailand, caloric contribution from rice as compared to other foods jumped from 50 percent to 80 percent after Green Revolution varieties were introduced there. By the end of the twentieth century, what had seemed at the time to be a scientific solution to quell world hunger had failed to address another food-related crisis for the poor: malnutrition.

Potrykus' Humanitarian Mission

It was this dilemma that Potrykus was determined to address. Ever since his early days as a plant scientist trying to transfer genes to make white petunias red, Potrykus had been mesmerized by the idea of using gene technology to solve global hunger problems. In 1973—not long after the success of the Green Revolution had been paraded—Potrykus began working with cereal crops in an attempt to further research that might have applications to food security. Although there had been rumblings that the techniques Potrykus had used with petunias might be applicable to food crops in developing countries, as Potrykus said at the time, "Obviously, to contribute, one would have to work with important crop plants and not only talk about them."

But manipulating genes in cereal crops turned out to be a good deal more difficult than making pink petunias. Despite the hurdles, Potrykus persisted with his work in the laboratory. Like Borlaug before him, he was appointed to establish a new research program focusing on bringing genetics to bear on food security, at the new Friederich Miescher Institute in Basel. By the 1980s, cereal crops were proving to be stubbornly resistant to the tools of genetic engineering, and each new technique that arose on the horizon continued to prove disappointing in attempts to transfer genes into cereal species. The importance of such work was not overlooked, however, and Potrykus was offered a prestigious position as a full professor at the renowned Swiss Federal Institute of Technology, where he continued with his attempts to transform food crops.

By 1990, Potrykus had narrowed his focus to rice, the staple crop of the majority of the world's poor, and had built a strong research team armed with the tools and persistence needed to make a breakthrough. The first breakthrough came when Karabi and Swapan Datta—a scientist couple Potrykus had recruited to his laboratory— found a technique that enabled the transformation of rice protoplasts, cells that have had their rigid cell walls stripped away to allow foreign DNA to enter the cell. By removing the outer cellulose-containing wall of the cell and incubating the resulting fragile protoplast with synthetic strips of DNA suspended in a chemical soup, the naked rice cells could be cajoled into sipping up some of the genes in their midst, with a select few of these rice cells going on to integrate the new genes into their DNA as if they were their own.

Using these techniques, the Dattas—working only nights and weekends on the project—were able to import into rice a gene native to the intestinal bacteria *E. coli* conferring resistance to the antibiotic Hygromycin B. The cells that had effectively picked up the bacterial DNA hitchhikers could easily be separated from those that hadn't by incubating them in the antibiotic solution, which killed off all but those cells expressing the incorporated *E. coli* resistance gene.

Certainly, antibiotic resistance is a useful tool and a common laboratory selection method in genetic engineering experiments with both plant and animal cells, but rice resistant to an antibiotic would make no contribution toward solving world hunger. Not unless, that is, the antibiotic resistance gene could be tagged onto another snippet of DNA encoding an important nutritional trait—say, for an enzyme that could catalyze the manufacture of high levels of B-carotene, the precursor to vitamin A in the body.

Then the antibiotic selection would really mean something, enabling the selective survival of plant cells that had a nutritionally important but otherwise invisible gene piggybacked onto the antibiotic resistance trait.

A Complicated Twist in Cajoling Rice to Make More Vitamin A

The commercial genetically modified crops engineered to express simple but economically valuable traits, such as the Bt crystal toxin gene or resistance to the herbicide glyphosate, were amenable to such an easy hookup. However, getting cells to express B-carotene was not just a one-gene operation. It turns out that the pathway that leads to the production of B-carotene requires the simultaneous expression of a number of different genes, many of which are not active in the endosperm, the starchy edible storage kernel of rice. Through an elegant series of experiments and related investigations, the necessary genes were identified by Potrykus and his crew. That crew now consisted not only of his own laboratory but that of Peter Beyer, an expert in daffodil terpenoids, the ring-shaped organic compounds that give spring flowers their brilliant yellow glow, and with it, a hefty dose of B-carotene.

Because they possess the necessary enzymatic machinery, plant cells like those of the daffodil naturally churn out high levels of vitamin A precursors. But not so with white rice. Rice, like all green plants, does produce B-carotene in its green and leafy tissues, but not in the endosperm. Further eroding the nutritional profile of rice is the fact that when it must be stored for long periods of time—as is often the case in regions where vitamin A deficiency is common—milling and polishing brown rice into white is required to prevent the rice from becoming rancid. Thus, in some of the most nutritionally compromised populations in the world, milled and polished white rice is the staple, rather than the nutritionally superior unmilled brown rice, with its nutrient-rich outer layers still intact. Even among the wealthy and elite, in many Asian countries the long history of reliance on processed, white rice has further entrenched the stigma of brown rice as a harbinger of poverty or wartime shortage, and has led to its limited presence at the dinner tables of all social classes.

For individuals whose diets out of necessity consist primarily—even solely—of white rice, various strategies have been devised to deliver back some of the missing ingredients in a feeble attempt to divert malnutrition. In recent years, the World Health Organization has cleverly attempted to sneak vitamin A supplementation into immunization programs for diseases like measles or polio. The group has also tried supplementation of vitamin A to new mothers, fortifying their breast milk, which passes on enriched doses of the vitamin to breast-feeding infants.

Yet high doses of vitamin A can be toxic and can cause fetal deformation if administered during pregnancy. In a joint immunization–vitamin A supplementation campaign sponsored by UNICEF in 2001 in the Indian state of Assam, fourteen children died and thousands more became ill, presumably from elevated doses of the supplement. The consequences to any fetus associated with overconsumption of vitamin A during pregnancy are so dire that in industrialized nations, prescription of the acne drug Accutane, a form of vitamin A, requires pharmacist and government monitoring of simultaneous contraceptive protection in patients.

Given these fearful prospects (not to mention the challenge of administering them), to Potrykus, none of the available solutions to vitamin A deficiency seemed anywhere near as effective or elegant as the vision of a rice plant genetically engineered to produce its own intact B-carotene. Well, perhaps not truly its own, but a version once encoded in the DNA of, say, a daffodil, and now faithfully replicated and assembled in the endosperm of an otherwise pale and B-carotene–less rice kernel.

But Potrykus was in the minority, even at the helm of his own research group, in believing that it would be possible to reprogram rice to impart a healthy, golden glow. Even though the B-carotene pathway had been identified, orchestrating the simultaneous expression of different foreign genes and keeping them from shunting off on biochemical sidetracks of their own remained a daunting task. Further complicating matters, the "biolistic" method of transforming plants by blasting them with gene-coated particles tended to make a mess of their own DNA, disrupting the genetic organization and inserting the new genes every which way in the genome. For the sake of stability and proving later, down the road, that these plants were genetically understood and predictable—which was a prerequisite for regulatory approval and consumer acceptance—it would be better to find a more tidy manner of inserting the foreign genes into the rice genome.

From Guns to Germs in the Quest for Golden Rice

Potrykus farmed this challenging project out to Xudong Le, a Chinese postdoctoral fellow in his lab who was eager to jump ship in order to take up a position in the United States. As such, Xudong was determined to solve this mighty dilemma in a year, and he was willing to take some scientific risks to make it happen.

The first change that Xudong made was to alter the delivery strategy for the DNA. He put down the biolistic gun and turned to a tried and true system of getting the necessary genes into rice via infection with a strain of bacteria called *Agrobacterium tumefaciens. Agrobacterium,* a resident of the living soil web, is naturally inclined to serve as a vector for DNA, spreading its own genes throughout the plant kingdom in the form of galls and tumors induced after infection at a wound site. When this property of *Agrobacterium* was first discovered in the early 1970s, it was hailed as a "natural form of genetic engineering," although the side effects of plant tumor formation made it initially unpalatable as a vector for GM foods. Soon, though, scientists had disarmed *Agrobacterium* of its own tumor-forming genes, replacing them with DNA designed to transfer a virtual cornucopia of traits from not only the bacterial kingdom but from other plants and even animals into an unsuspecting plant.

Agrobacterium offers a number of advantages over the biolistic method of propelling foreign DNA into a plant cell. For one thing, *Agrobacterium* often does a better job of integrating individual copies of desired genes intact, rather than scattering them in multiple copies or fragments throughout the genome of the new host. This enables more stable and predictable expression of the foreign gene in its new location, which is a priority when it comes to genetically modified food. Second, *Agrobacterium* can be used to infect a single cell, which can then be treated with various hormones and chemicals to regenerate an entire, genetically homogeneous new plant with the introduced gene intact in each of its cells. Shooting genes at random into a suspension of cells, on the other hand, leads to a heterogeneous population of plant cells with different numbers and positions of genes inserted in each one. When new plants are regenerated using this method, the result is often a mosaic, with the desired gene expressed differently in different parts of the plant.

There was only one problem with Xudong's eagerness in using *Agrobacterium* in an attempt to create Golden Rice. In nature, *Agrobacterium* infects only dicotyledonous plants, or those that sprout two seed leaves; have a reticulated, netlike pattern of leaf veins; and organize their vein-like vascular bundles into rings at the center of their stems. Luckily for the biotechnology industry, several of the more lucrative commercial crops such as soybeans, cotton, and canola are hardy members of the dicot tribe, and had thus been proven tractable by the *Agrobacterium* means.

Monocotyledenous plants, on the other hand—those whose sprouts spring forth with only a single seed leaf and scatter their vascular bundles at random throughout their stems—had proven recalcitrant to infection with *Agrobacterium*, both in nature and in the laboratory. Because rice, like most grasses and cereal crops, is a monocot, the proposal to use *Agrobacterium* in the creation of Golden Rice made it seem even less likely to succeed.

Undaunted and armed with some skillful laboratory tricks, Xudong managed to wrestle his rice plants into submission, infecting a good number of them with a form of *Agrobacterium* loaded up with genes borrowed from daffodils and from *Erwinia,* a bacterial plant pathogen. By the spring, he had managed to establish fifty different transformed plant lines, some of which when polished gave off a fateful yellow glow suggesting success. Against the odds, Potrykus and his team had managed to wheedle the rice endosperm into expressing the novel genes needed to produce B-carotene in its cells, and the first prototype of Golden Rice was born.

Golden Rice Remains in Waiting

This was in 1999, and in the years since, Golden Rice has encountered a vast number of regulatory hurdles that have as yet prevented it from making its anticipated humanitarian debut. Xudong's initial Golden Rice produced only around 1.6 micrograms of B-carotene per gram of rice, provoking critics to state that people would need to eat between 3 and 5 pounds of such rice each day to meet their recommended daily allowance of vitamin A. Since the initial strains were unveiled in Potrykus's lab, new and improved varieties of Golden Rice have been produced. These new strains rely on genes from corn and from rice itself instead of daffodils, and they manage to churn out nearly twenty times more B-carotene than the original precursor Golden Rice. The B-carotene genes have been transferred from Japonica varieties of rice—the short-grained varieties in which Golden Rice was initially developed—to the longer-grained Basmati-like Indica varieties commonly cultivated and consumed by much of the world's poor.

With all of these improvements, why does Golden Rice still sit on the shelf, while commercially available transgenic crops like Bt corn and cotton continue to increase in global acreage? Will Golden Rice and GM traits be the salvation of the poor and save GM from the erosion of its opposition? Or is Golden Rice, as its critics contend, merely a humanitarian smokescreen, a Trojan horse that the biotechnology industry intends to ride through poor and unsuspecting nations, spreading the seeds of species transgression and patents on life in its path?

It is here, at the juncture of these debates, that the opponents of genetically modified food and the developers of Golden Rice come curiously and
coincidentally close to seeing eye to eye. Not that they would ever admit to
it, of course, although Potrykus initially welcomed his critics and even
agreed with them. As he wrote to one of his most vocal critics in February
2001 in response to criticisms of the limited amount of B-carotene the original strains produced: "I am happy to acknowledge that Greenpeace is arguing on a rational basis . . . I also acknowledge that Greenpeace has
identified a weak point in the strategy of using Golden Rice for reducing
Vitamin A deficiency." But this was merely a temporary deviation from an
ongoing battle between Potrykus and his opponents. A year earlier, he had
accused his critics of taking advantage of a much needed solution to world
hunger to promote their biased anti-biotech agenda. "What these radical
opponents are doing is 'Brunnenvergiftung' [well-poisoning] to the disad-
vantage of the poor," concluded a memo Potrykus had posted on a pro-
biotech website.

Whether or not the well had been poisoned did not matter much at
this point, because there were far more serious obstacles in bringing the
bucket to those who needed to drink, so to speak. Until numerous patent
and regulatory hurdles could be cleared, Golden Rice would indeed remain
at the bottom of the well, inaccessible to those who needed it most, whether
they were the poor and hungry or merely part of the biotechnology industry
badly in need of a brushed-up image.

Patents, Profits and Philanthropy in the Golden Rice Dilemma

The ethical dilemma of patenting genes from living organisms, particularly
in agriculturally important food crops, had long been one of the opposition's
strongest arguments against GM food. Several high-profile lawsuits had only

fueled such arguments. Family farmers were clobbered by gene-giant corporations like Monsanto when GM plants were found on their farms without
the required purchase agreements.

Take Mitchell Scruggs, for example, a fifty-four-year-old cotton farmer
whose family had plowed their Mississippi farmland for more than a century. Scruggs was one of a large handful of family farmers sued by Monsanto
for reusing seed they had purchased from the company. In pursuing their
case against him, the company sent numerous investigators to stake out his
property, and hired planes from a local airport to conduct overhead surveillance. They even went so far as to purchase an empty lot across the street
from his property to serve as a base for their operations. Monsanto's intellectual property protection rules prevent farmers from saving or replanting
seed from their harvests, which is an age-old conventional practice in agriculture. Such rules mean that farmers must return to the source each year,
paying in full each time, to get seeds for planting. And not just any source;
the sole purveyor of seed for a particular GM trait or variety is the company
that holds the patent on it. While most patents are painstakingly descriptive
and are often narrow in scope, when it comes to GM seed, patent criteria
have often been applied with a broad and liberal brush. For example, between 1994 and 2007, Monsanto held an exclusive European patent on all
species of GM soybeans, regardless of the trait or technique used.

The issue of patents, which were nowhere to be found in Borlaug's
Green Revolution circa 1970, dropped a distasteful dilemma into the humanitarian angle of the Golden Rice debate. At the time that Golden Rice
made its debut, and continuing up to the present time, all commercially
available GM seeds have been protected by patents, and nearly all of the
techniques used in making them have as well. How will such proprietary
property make its way to poor farmers, without further adding to their economic burden through costly annual royalty payments? Potrykus initially
brushed this issue aside by assuming that he and his co-discoverer Peter

Beyer could patent Golden Rice themselves and provide it free of charge. But given the path of patenting that had been laid by the GM giants, this task was not a simple matter.

Potrykus found this out the hard way. When the first successful prototype of Golden Rice had finally yielded its elusive yellow kernel, the Rockefeller Foundation commissioned an audit of intellectual property rights that had been used in its making. The result was shocking. Thirty-two companies owned seventy different intellectual and technical property rights, otherwise known as patents, on the various laboratory techniques and procedures that Potrykus and his team had used in making Golden Rice. In order for Golden Rice to be distributed commercially, each of these would need to be negotiated. As Potrykus said at the time, "It seemed to me unacceptable, even immoral, that an achievement based on research in a public institution and with exclusively public funding, and designed for a humanitarian purpose, was in the hands of those who had patented enabling technology early enough or had sneaked in a material transfer agreement in the context of doing an earlier experiment. It turned out that whatever public research one was doing, it was all in the hands of industry (and some universities)."

This argument sounded oddly reminiscent of anti-biotech crusaders' now recurrent insistence that patents on living organisms and the monopolies on life they might lead to would stymie research. Such patents, they contended, would prevent, rather than promote, safe and humanitarian applications of biotechnology, potentially threatening the environment in the meantime.

Beyond the practical hindrances that gene patents might present, many non-governmental organizations (NGOs) active in the fight against biotech also played the ethics card. The Greenpeace website alleged, for example, that "Life is not an industrial commodity. When we force life forms and our world's food supply to conform to human economic models rather than their natural ones, we do so at our own peril." Indian scientist and anti-biotechnology crusader Vandana Shiva was even more harsh in decrying the destructive potential that acts allowing patents on food crops posed in her native India. She claimed that allowing such patents "could forever destroy the biodiversity of our seeds and crops, and rob farmers of all freedoms, establishing a seed dictatorship."

Granted, this type of argument was completely out of line with Potrykus's movement to promote Golden Rice, and it did not take long for him to dramatically shift his tone, especially when AstraZeneca, one of the major patent holders of the Golden Rice technologies, offered to negotiate a patent deal that would allow free access for poor farmers. Now, Potrykus claimed, he had done "a bit of further thinking and became aware that 'Golden Rice' development was only possible because there was patenting. Much of the technology I had been using was publicly known because the inventors could protect their right."

But Golden Rice was still struggling to make it to market. Potrykus turned to government regulatory policies as a scapegoat. Now, it was not the patents themselves that posed a problem but the extensive regulation required for approval of a new GM product. With this turn of reasoning, GM critics could be the enemies again, since it was their insistence upon safety testing and risk assessment—a draconian insistence in Potrykus's view— that drove such regulations and was now the barrier to Golden Rice's lifesaving potential.

Regardless of who or what is at fault, it is clear that two characteristics of Golden Rice distinguish it from the earlier progenitor Green Revolution varieties and have hindered the pace at which its humanitarian potential might be tested. The fact that intellectual property rights continue to govern its distribution—allowing free access for farmers earning less than $10,000 per year but still imposing royalty fees and licensing agreements for any other commercial use—certainly stands in the way of releasing the seed freely in whatever fields it might be sown. And, as Potrykus has correctly pointed out, the transgenic nature of Golden Rice has subjected it to government regulatory policies and triggered a slew of tests and permits required to prove its health and environmental safety. Many of the countries that are the intended targets of the technology don't have biotechnology-oriented policies in place to begin with, presenting even further challenges to obtaining the necessary approval.

Even once these hurdles are surpassed, Golden Rice is unlikely to fill the gaps that remain in the decades-long wake of the Green Revolution that have only widened since the 1970s. Much of the success of Borlaug's work impacted countries in Asia and Latin America, leaving large swathes of African nations entrenched in serious situations of food insecurity. While Golden Rice may hold the potential to provide a technological fix for malnutrition in rice-producing and consuming nations, it is likely to once again leave behind some of the most vulnerable and overlooked populations, especially those in Africa. While the example of Golden Rice provides an exception to the rule of bottom-line-driven biotech crops, it is far from a miracle cure and has perhaps been predisposed to being overblown by big companies eager to overlook the criticisms of their currently existing GM regimes. Thus, it should come as no surprise that any attempt at winning the public's hearts and filling their shopping carts in return with the humanitarian message of GM food has a long way to go, even in nations hungry for solutions to food security crises.

the current week's question: "Why do you think some African nations are rejecting genetically modified crops? Because of health concerns or politics?" the anchor bubbled out, apparently glad to turn away from the bad news at home and focus on someone else's more dire misfortune.

War and Famine

It wasn't exactly breaking news. Several months earlier, as a drought-plagued African harvest season had come to a close and food had been in short supply there, a slew of sub-Saharan African nations had raised objections to shipments of genetically modified food aid headed for their shores. While most countries had backed down, Zambia in particular had held out. Its president, Levy Mwanawasa, proclaimed such food aid to be "poison" that he would rather die than consume.

The United States seemed to be poised on the brink of a war in Iraq that it might launch without the support of its allies, but the food aid rejection still stung enough to be put on the nightly news. Perhaps, as others would later say, it wasn't just a little tragic relief, a turn of the camera to the distraction of famine on distant shores. Perhaps it was intentional, and all just part of one big, messy story weaving together politics, food, and a quick biotech fix.

When the *MoneyLine* broadcast finally made its way to the issue, it was famine in Ethiopia that made the headlines and the photo op, even though the nations that had rejected genetically modified food aid—Zambia, Zimbabwe, and Mozambique—were found far to the south. Besides, in Ethiopia, as the newscaster accurately reported, "There may be hunger. There's also plenty of grain on sale for those who have money to buy it. Even vegetables. Few here can afford them." Thus, despite the fact that food was available to some in Ethiopia, food aid was still needed there.

Seamlessly flowing into coverage of the GM food aid situation, Henry

Miller, a fellow at Stanford's Hoover Institute, offered his judgment on the question of who was to blame for African countries' unwillingness to accept such goods: "It's ironic that one of the villains of the piece is the United Nations itself along with the European Union . . . It's a moral outrage, it is. And it's partly concerns about trade issues and partly just power politics played by the EU."

Previously, Andrew Natsios, a representative of the U.S. Agency for International Development (USAID), had been even more harsh in his accusation of the groups he considered to be behind the refusal. "They can play these games with Europeans, who have full stomachs, but it is revolting and despicable to see them do so when the lives of Africans are at stake," he said, later adding, "The Bush administration is not going to sit there and let these groups kill millions of poor people in southern Africa through their ideological campaign."

Ten days earlier, Robert Zoellick, the U.S. Trade Representative, had called European policymakers "Luddites" and had pinned African rejection of GM food aid on them. And later, in May 2003— when non-GM food aid had been found and delivered to the nations in question and the matter should have been long forgotten—President George W. Bush would bring up the issue up once again. In a commencement address at the U.S. Coast Guard Academy in Connecticut, Bush—after paragraph upon paragraph regaling America's elimination of aggressive terrorist threats in the name of global freedom— concluded his speech by scolding European governments for failing to support his new Initiative to End Hunger in Africa: "We can also greatly reduce the long-term problem of hunger in Africa by applying the latest developments of science . . . Yet our partners in Europe are impeding this effort. They have blocked all new bio-crops because of unfounded, unscientific fears. This has caused many African nations to avoid investing in biotechnologies, for fear their products will be shut out of European markets."

So there it was, laid out on the table, like a victory lap following the self-congratulatory "end to combat operations in Iraq" that Bush had issued just three weeks before. *You are either with us or against us.* The global politics of genetically modified food left no room for neutrality, just like the "War on Terror" campaign that had been under way for two years. A Greenpeace campaign cast the standoff in even harsher terms: "Eat this or die: The poison politics of food aid."

The political passion and energy that U.S. government leaders directed at Zambia's quiet but steady refusal to accept genetically modified U.S. food aid seemed out of proportion with the issues of the day. Some might wonder why a little fray over food aid figured so prominently on the agenda of a government fresh from a war that might not really have ended yet, and how the specter of famine in far-off Africa could have suddenly risen to the forefront of the nation's priorities. Certainly, hunger in Africa had swept back and forth across the continent throughout the 1990s, in sinusoidal synchrony with climate change and political conflict there. While the severity of the food shortages that had been experienced in Africa in the fall of 2002 were significant, they were not new. Nor was the influx of aid in the form of food, which had been flowing to foreign shores with regularity for more than half a century.

Food Aid for Peace, Power and Profit

Hunger in Africa, and its link to the pipeline of U.S. food aid, had long been far more than merely a heartfelt humanitarian concern. In 1954, President Dwight D. Eisenhower had signed into effect Public Law 480, the Agricultural Trade Development Assistance Act, with its stated goal being to "lay the basis for a permanent expansion of our exports of agricultural products with lasting benefits to ourselves and peoples of other lands." Eisenhower had also urged the United Nations to devise "a workable scheme . . . for providing food aid through the UN system."

Shortly after instituting the U.S. Agency for International Development in 1961, President John F. Kennedy renamed the PL 480 Program "Food for Peace," proclaiming that "Food is strength, and food is peace, and food is freedom, and food is helping to people around the world whose good will and friendship we want." Or the other way around. As one Zambian official put it the summer following the GM food aid debacle, "Small nations like Zambia are in a difficult position; we walk a fine line, a very fine line. America is the dominant power right now, and its power works through the IMF [International Monetary Fund], the World Bank, the UN even, so small nations like Zambia don't want to take on the might of the U.S."

USAID was even more clear. A statement on its website at the time of the GM food aid debate proclaimed, "The principal beneficiary of America's foreign assistance programs has always been the United States. Foreign assistance programs have helped create major markets for agricultural goods, created new markets for American industrial exports and meant hundreds of thousands of jobs for Americans." In a press release for the agency's Collaborative Agricultural Biotechnology Initiative, its stated goal was to "help developing countries access and manage the tools of modern biotechnology." Agenda item number three was quite clear: "Local private sector development will help to deliver new technology and integrate it into local agri-food systems."

Monetization and the Ethics of Food Aid

Recently, some longstanding and respected participants in the food aid business have turned a critical eye on such policies. In 2007, CARE—one of the world's largest aid organizations—snubbed its nose at $45 million in federal financing for programs it proclaimed unethical. In these programs, the U.S. government purchases surplus food commodities from American agribusinesses, since under U.S. rules no cash donations are allowed (despite the fact that all other countries sending food aid provide cash to the receiving country). The U.S. government then pays to have the surplus food sent to the receiving country on a U.S. carrier. Cargo preference laws stipulate that a minimum of 75 percent of food aid must be shipped by U.S. flag carriers, and a U.S. Government Accountability Office study in 2007 found that 65 percent of the cost of food aid was gobbled up in transportation costs, rather than in food itself. The food is then "donated" to aid groups as "an indirect form of financing."

But that does not mean that hungry recipients will necessarily receive that food for free. In an odd twist that smells of food laundering, in many cases the donated food is sold by the aid groups on the open market in the receiving country, and profits from that sale are then used to finance aid programs. Groups like CARE argue that this practice, called monetization, tends to benefit U.S. interests more than the poor countries. Such strategies tend to drive down market prices on the receiving end, further debilitating these countries' economies and putting an ethical kink in the concept of food aid itself. Anti-GM activists as well as broader aid organizations suggest that it is for this reason that the delivery of in-kind food aid functions as a Trojan horse for U.S. interests—one of which is planting the seeds of pro-GM sentiment both literally and figuratively in foreign countries.

In this regard, food aid and the food shortages that drive it are big business in the United States. By the time the GM food fray hit in 2002, the U.S. government was equipped with high-tech famine trackers, supported by a number of government agencies, that allowed precise prediction of lo-cales that were likely to experience food shortages in the near future. The management of this network—known as FEWS NET, for Famine Early Warning Systems Network—is run by Chemonics International, a for-profit organization that derives 90 percent of its business from USAID contracts.

In addition to running FEWS NET, the company has been the recipient of a $25 million contract to help to bolster the economy in Iraq, on top of a $60 million subcontract for a USAID-funded Iraq local governance contract. The U.S. Geological Survey, one of the principal federal agencies involved with FEWS NET, states on its website that: "Chemonics maintains a staff of people in various African countries who are the 'eyes and ears' of the FEWS NET project. Their job is to collect and analyze various types of data, take field trips and otherwise keep abreast of conditions in individual countries. The field staff also send their data and analyses to the FEWS NET office in Washington, D.C., which is also managed by Chemonics."

On the Ground in Zambia

And so, it seems, tracking famine is a major U.S. government pastime, a seemingly generous and altruistic exertion of its might in the name of the global humanitarian good. And no wonder: From the American vantage point, food aid has become part and parcel of the global economy that holds the United States, with its subsidized agricultural surplus, high above the heads of most other nations—most especially small, economically struggling nations like Zambia, which most Americans had never even heard of in 2003, let alone were able to identify on a map.

Which may explain why CNN's news coverage featuring starving Ethiopians with extended bellies and shrunken silhouettes hit such a nerve with expatriate Zambians watching the nightly news from their American posts. "Those aren't starving Zambians, those are Ethiopians!" the relative of one Zambian minister phoned in to report after viewing a major news network's coverage of the Zambian food aid rejection. Apparently the network had spliced in clips of generic "starving Africans" to accompany its Zambia news piece and had drawn on stock footage filmed in Ethiopia to dramatize the segment rather than relying on actual real-time reporting. It was true that in Zambia there were shortages of the staple corn (or maize, as it is known in Zambia). However, the country did not have the all-out famine that swelled bellies and bared ribs and played so well on nightly newscasts and right into most Americans' stereotypes of African famine.

Things were a bit different on the ground in Zambia. There was hunger and certainly vast gaps in the nation's supply of its staple maize. But hunger means different things to people in different contexts. While the random bouts of hunger that all too unfortunately accompany poverty the world over—from small towns in rural Mississippi to the expanse of sub-Saharan Africa—was indeed widespread in Zambia during the winter of 2002 and into 2003, the situation was more complex and less dire than superficial reports had implied. There was food in Zambia—good food, in fact, and plenty of it. One of the greatest problems was that such homegrown commodities could not be depended upon to be in the right place at the right time. In the fertile north, food was plentiful and stomachs for the most part were full. But in more remote and drought-stricken locales to the south, food stores had hit rock bottom and health and morale had plummeted. The lack of roads and transportation options in Zambia stood as an often insurmountable barrier to the ability to move existing food around.

Food aid supplied through the UN's World Food Program, on the other hand, came complete with the trucks, helicopters, and mud-crawling convoys needed to bring donated food to the hungry in even the farthest reaches. As the food aid operation went into force, one Zambian government health worker shook her head. "It's a shame they can't just send the equipment and let us use the food that's here. Food aid makes beggars of us all," she said.

That was one take on why the picture of hunger so hauntingly painted by the Western news media failed to fully capture the experience on the ground. Another view was offered by a professor of agriculture at the University of Zambia. Sitting behind an elegant wooden desk overlooking fields turned brown from lack of rain, he explained: "In the issue of food aid, you must ask the right questions. If you ask, 'Have you eaten today?' one will answer 'no' unless there has been nshima (a cornmeal product that is a staple in Zambia) that day. Even if there has been sweet potato or other foods . . . when we eat rice, that is not a meal. For me today, I have brought a sandwich, but I feel that I have not eaten until I have my nshima. So you see, the questions must be correctly researched."

Any American tourist landing in Zambia and sitting down for dinner might immediately agree with the professor. Maize makes up 60 percent of the calories in the typical Zambian diet, primarily in the form of nshima, which is a heavy, grit-like mound accompanied by a sauce with meat or vegetables that is eaten two to three times per day. After adjusting to such a diet, which is served across classes and can even be had at one of the few high-end restaurants in Zambia, it is indeed difficult to feel full when eating anything else.

The Colonial Legacy of Maize in the Zambian Diet

While many Zambians of varying educational backgrounds will tell you that maize is indigenous to their culture and country, an exploration of the history of how maize came to figure so prominently in the Zambian diet tells a different story. Maize was brought to the African continent by the Portuguese, during the days of the slave trade in the sixteenth century. It quickly took root, so much so that the popular history and culture of many African nations incorporated its presence as a native foodstuff.

In Zambia, the earliest documented agricultural policy in the country's history was a mandate that white farmers grow maize to feed mine workers. In 1936, colonial rulers implemented a Maize Control Board (MCB) to regulate the production and pricing of the crop in the country. From there, maize cultivation spread, until by the 1980s maize was the chief target of agricultural efforts in the country and made up more than half the caloric intake of its citizens. Economic incentives and neglect of the agricultural sector in favor of mining and industry quickly swept the traditional polyculture of cassava, sorghum, and millet aside in the wake of "King Corn," as some historians have called it.

But as demonstrated by a series of devastating droughts in the 1970s and 1980s and the resulting low point in maize production, dependence on such a single crop made Zambia vulnerable, and something needed to be done. The wheels of reform slowly ground into motion, but it would take another two decades and a revisiting of the dangers of drought-induced dependence before Zambian officials and agriculturalists would formally recognize that "Diversification is Key."

Conservation Farming: A New Way Forward?

Mundia Sikatana—the minister of agriculture at the time of the GM food crisis, and later minister of foreign affairs—was a staunch proponent of agricultural diversity and independence from the legacy of food aid. Sikatana was a strong supporter of a new campaign in Zambia known as "conservation farming." The movement toward conservation farming techniques had begun in 1996, supported by a broad coalition of stakeholders including not only Sikatana and the government ministry but several donor organizations and members of the private sector as well.

Conservation farming follows in the footsteps of the growing global organic movement, yet it is not so rigid as to disallow chemical fertilizer inputs. The basic premise of conservation farming is based on reducing tillage of land in the dry season, composting crop remains from the growing season back into the soil rather than burning those potential nutrients up, carefully targeting inputs and seeding, and a crop rotation system involving leguminous nitrogen fixing crops in the off-season. All of these techniques together aim to increase yield and retain soil moisture, particularly in times of drought, which is a serious issue in a nation where most small-scale agriculture proceeds without any formal system of irrigation at all.

It might seem far-fetched if not outright ostentatious to suggest that a bit of composted garden waste and a pullback on tilling could solve the enduring hunger problems that cycle in and out of many Zambians' lives, but conservation farming is beginning to show effects. In the midst of the food aid debacle, a survey of 125 farms located in the central and drought-plagued southern regions of Zambia indicated that on average, small-scale farmers employing conservation farming produced 1.5 tons more maize per hectare than their compatriots utilizing the conventional ox-plow tillage methods. Of this bounty, approximately 1.1 tons per hectare were estimated to be a direct result of conservation farming techniques, while the remaining 400 kilograms per hectare stemmed from increases in fertilizer, lime, and high-yield varieties of seed.

Sikatana's own observations confirmed that all this was achieved without the need for GM technology and in some cases even any chemical inputs. "I know of small-scale farmers, women, that are not using fertilizers, that have enhanced their yield without any use of biotechnology," he said. "They have harvested a record because we are supervising them and they have embraced conservation farming, and they are very proud."

Moving Beyond Maize

Beyond changing the methods by which Zambia's subsistence farmers grow maize, Sikatana also emphasized the need to diversify the nation's diet, weaning away from maize and expanding the demand for more drought-resilient crops like cassava, millet, and sorghum. Cassava had tagged along behind maize as a colonial crop imported by the Portuguese and had long been a backup crop to supplement maize production in times of drought. Indeed, colonial legislation ordered Zambian farmers to include a small plot of cassava in all plantings, under the threat of penal sanctions.

In interviews conducted by agricultural researchers Steven Haggblade and Ballard Zulu, "Old farmers volunteered that District Officers would come to inspect their cassava fields armed only with a stone. Standing in one corner of the cassava field, the inspecting officer would throw the stone as far as he could. If he could launch it out of the field, the field was too small." For this reason, it is hypothesized that cassava fell out of favor as a mandated colonial crop, though it is now beginning to make a comeback, ironically as a means to escape the neocolonial dilemma of dependence on food aid. Since the initiation of cassava reform efforts, the landscape of agricultural allocation to maize has shifted somewhat. Between 1982 and 1999, the total percentage of cropped area consigned to maize fell from 80 percent to close to 60 percent. In the same time period, root and tuber cultivation increased from 20 percent to 30 percent of total area.

Dietary diversification in Zambia also extends beyond cassava and is again an example of asking the right questions. An informal faculty focus group at the University of Zambia suggested that the true scope of the Zambian diet, while still relatively centered around maize in the form of nshima, remains invisible to the Western eye. When it comes to food aid, "We need to look at what is appropriate for Zambians. Here people eat five or six different types of caterpillar—we have no need for a gene to kill off the caterpillars!" said one determined professor, referring to the cultural inappropriateness of the Bt trait engineered into the food aid. "Also with labor," he continued. "Here we eat the weeds as greens, and people only work about 500 hours per year, so they do not see labor associated with weeding as a problem," he concluded, referring to Roundup Ready and other herbicide resistance traits pushed largely for their weed-reducing properties. While such views may be criticized as the privilege of the elite, a visit to many markets proves this to be true, with baskets of dried caterpillars laid out in the sun, and many a foodstuff that would rapidly be deemed a pest proffered as a delicacy.

Diversification is Key

In the wake of Zambia's food aid fiasco in 2002, the theme of agricultural diversity appeared to have taken hold with a vengeance. The country's annual Agricultural and Commercial show—a national jamboree of sorts—

held the following year had as its theme "Diversification is Key!" In sharp contrast to the dismal tenor of the gathering the previous year, at this over-crowded event on the sprawling national fairgrounds, smiling faces showed off chickens and honey, kente cloth and schoolbooks. Over by the agricultural hall, a display of irrigation equipment glared out bright and green against the dry winter backdrop. These were machines that most could not afford, put on show as something to marvel at rather than really take home under your arm.

Inside the hall, representatives from various universities and extension centers hawked their newest data and innovations. Seeds of many shapes and colors shone brightly out from displays reiterating the theme of diversification. Fruits and vegetables colorfully decorated tables where students and extension agents eagerly handed out business cards advertising their expertise and programs. In the face of the foreign news reports from the past summer, this place seemed too happy to be true, a virtual bounty of food and agricultural productivity with only a passing reference to the struggles of drought and certainly no hint or insinuation of famine.

And there, over in the corner near the door, stood a display that purported to be able to detect whether GM content was present in a sample of maize. Telltale bands on a wavy gel electrophoresis test singled out those lanes that held contaminated samples from those that did not. Having the power to detect whether or not a local stock of seed had been commingled with the imported GM variety was seen as an absolute necessity before Zambia could even think of accepting GM food on its shores. One of the strongest arguments the government had put forth was the lack of any regulatory policy in place by which to make decisions about the presence of GM crops in the country. Paralleling that argument, there had been a vocal cry for an initiative to develop the nation's biotechnology expertise and ability, presumably to allow independent detection as well as decision making about the looming biotech future.

From Food Aid to Field Contamination

Even if the presence of transgenic crops could be detected—as Zambia National Farmer's Union representative Lovemore Simwanda would carefully explain they had—the question of what could be done about them still remained. One of the main arguments for rejecting GM food aid was the fear that they would contaminate the country's own non-GM seedstock.

In Zambia, farmers by necessity save seed from year to year, holding back up to one third of their harvest to supply the stock for the next year's crops. Because hybrid corn does not continue to "breed true" and can over time lose its treasured traits, farmers in Zambia do return to their own local seed companies, such as Zamseed, every now and then to replenish their stock. Should GM corn provided in aid be saved and planted by farmers—as anyone assessing agricultural conditions in Zambia would certainly agree that it would—the potential to contaminate commercial fields and even the national seed companies' stocks would loom large, spoiling the market opportunity for export to Europe. Indeed, Zambia's largest employer, the agricultural company Agriflora, was a staunch opponent of GM for precisely this reason. Agriflora emblazoned its growing fields with faded tin signs clearly admonishing "Keep Zambia GM Free."

It was for this reason that the United States had accused European policymakers of causing Zambians to starve. According to the argument, if European markets would not accept GM food, and Zambia had a major economic stake in supplying the European market, then any GM contamination would close Zambia out of its arguably most lucrative market. (In fact, Zambia did have a major economic stake in the European market: A quick check of London supermarket chains that summer turned up petit pois, mange touts, and haricots verts, all tiny-sized gourmet organic vegetables, marked with Zambia as their country of origin.) And it was likely to be true that Zambia would be shut out of this market. Although the only commodity that might be at stake was corn, since that was the only GM species present in the food aid, it was still not trivial. Zambia was rumored to be a significant supplier of baby corn to the United Kingdom, the largest baby corn importer in Europe.

But was this risk alone enough to sway Zambian policymakers away from accepting aid? Although important, casting Zambia as an economic victim of draconian European food preferences was only a small part of the story. While European sentiment and distaste for GM food certainly influenced Zambia's decision to stand firm in the face of pressure to accept the food aid, the country had approached the question quite thoroughly and had come to the decision on its own.

Zambia's Fact Finding Mission

When the food aid issue arose rather unexpectedly in the summer of 2002, the country realized that they were practically and scientifically unprepared to stand behind a decision either way. The government formed a fact-finding delegation, made up of experts in science, health, nutrition, and agriculture, to gather information relating to the issue of GM food. The mission's first stop was Europe, where its members met with experts and advocates on both sides of the issue.

Not to be outdone, the delegation was quickly wooed by the United States and invited to visit with scientists and policymakers on the other side of the Atlantic, culminating in an international Ministerial Conference and Expo on Agricultural Science and Technology in Sacramento, California, jointly sponsored by USAID and the U.S. Department of State. As one Zambian government official present at the meetings later recalled, "Our relationship to the United States and the impressions that we got at Sacramento had led us to believe that we have to tread cautiously, very, very cautiously. I had the impression that America is like a wounded buffalo."

Members of the task force returned to Zambia to present their findings to the public. An *indaba* was called, an indigenous form of community decision making tracing its roots to Zulu culture. This come-one, come-all gathering was held at the Mulungushi conference center, a shiny exhibition venue on the outskirts of the capital Lusaka. There, attendees heard reports from the fact-finding mission and expressed opinions of their own. Overwhelmingly, the outcome was in favor of continued rejection of GM food. A few lone voices, some later accused of having covert ties to the U.S. biotech industry, tried to speak up in favor of the scientific certainty of GM's safety and salvation. But they were quickly hushed by the din of outright opposition. As one pro-GM health scientist described it, "I trooped off to the debate to go and hear what the people were saying . . . I was given the opportunity to speak, and I SPOKE!" He gestured with his hand and gave a quick laugh. "I wasn't very welcome in the meeting, no, not at all! My views were at variance with everyone else's . . . I said, well yeah, the agriculturalists had reason to be concerned, and so also the environmentalists, but as to us in health, really there was no reason for concerns."

Another healthcare worker who was a member of the advisory team, did not brush off the human health concerns so lightly. As a participant in the fact-finding delegation, he felt dissatisfied, even alarmed, by some of the answers he was given by U.S. officials. "We were told there were studies done in rats, but not in humans. They told us it was unethical to do tests in humans, but then is it logical that they want to give it to our people without a test? If it is unethical to do tests, then isn't it unethical to give it to us without testing, as well?"

Indeed, one of the resounding arguments presented by the U.S. government in defense of the safety of GM foods was the refrain, "But Americans have been eating GM food for seven years, and nothing has happened to them!" Of course, that was neither a scientific nor a particularly reassuring response. As one member of the fact-finding mission pointed out, "Americans are not particularly healthy. Look at diabetes, cancer, look at obesity! We do not want to increase these things, so why should we eat what they do?"

Although a direct link was unlikely between consuming GM corn and, say, contracting diabetes or cancer or indeed the epidemic of obesity bursting the seams of America, he certainly had a point with his guinea pig argument. Individuals dependent on food aid were likely to consume maize as their sole foodstuff, being unable to afford anything else to eat. They were also likely to be women, children, or the elderly, who are statistically the poorest of the poor, not just in Zambia but worldwide. And in Zambia, where close to one-third of the population was HIV-positive, they were also likely to be suffering a massive breakdown of their immune system as a result of the debilitating symptoms of AIDS.

It was this population, characteristic of those most likely to be dependent on food aid, who would be consuming large amounts of unprocessed GM maize. This group was not much like the oft referred to Americans who consume corn primarily in the form of highly processed byproducts like high-fructose corn syrup, maltodextrin, or monosodium glutamate. Given these conditions, the GM food aid strategy did smell suspiciously unscrupulous. Although as U.S. officials reiterated, health studies in humans would be unethical, it was hard to imagine a better study design for an experiment formulated to detect any human health impacts that might arise from eating GM maize than the one proposed to take place in Zambia that summer.

Besides, this was not the first time that a gift horse had kicked Zambia in the mouth. Back in the 1980s, in the wake of the nuclear power plant catastrophe at Chernobyl, a donor country had delivered a shipment of food aid in the form of tinned meat to Zambia's poor. Tests of the contents indicated that the meat harbored invisible but unsafe levels of radiation and the aid was recalled, but not before the hungry hordes had gotten their hands on what to their eyes, ears, and taste buds appeared to be a nutritious and tasty meal. The Zambian government was forced to bury the remains of that contaminated consignment in deep graves to prevent further exposure of the population.

This experience, admittedly along with the observation of Europe's vehement refusal of GM foodstuff, sealed the fate of GM food aid for Zambian officials. And so the decision was affirmed and GM food aid was locked out, kept from the public in the large storage silos where it had been moored since it had arrived in ships on African shores and been shuttled inland to the site of the supposed famine.

GM Maize Takes Root

But that did not mean that GM maize was not to be found in the country. The summer after shipments of GM food aid had arrived, the Zambia National Farmer's Union had amassed evidence that the contraband GM corn had been planted in the country. Standing in the Zambia National Farmer's Union office, Lovemore Simwanda pulled out ears of corn from the 2003 harvest from a box that had been stored there. Whereas Zambian varieties are laden with white or lavender kernels, these cobs had crooked sets of yellowed kernels interspersed with the white and purple. The yellow color of the kernels—common to corn grown in the United States—had sparked suspicion that the GM maize had taken root. Molecular tests had indicated as much, and scientists from Cornell University, in the United States, were rumored to be using GPS to help determine the site and extent of the spread.

It was not only in the offices of the Zambia National Farmer's Union that the outlawed foodstuff could be found. In the settlements of Misisi township, on the outskirts of Lusaka, a woman standing in the doorway of her cinder block shanty extended a fist to reveal a handful of yellow corn she claimed had been smuggled from stores of the forbidden food aid. When asked where she had gotten it, she just smiled, flashing a gap-toothed grin. It was here where the specter of hunger was the greatest, where residents

picked through trash piles, and cinder block squats 8 feet square slept families of ten or more each night. The settlement was haunted by AIDS, such
as when a gaunt young woman passed, clutching a shriveled infant, begging
for money to purchase formula that would protect her baby from her virus-
tainted breast milk.

Food to Be Had, Despite the Drought

But nonetheless, despite these tragic signs of poverty and squalor, food was
present if not plentiful. Behind a small hovel, a woman butchered a chicken,
red blood seeping from its inverted head as its legs twitched into silence.
Kerchief-crowned women stirred boiling katchasu, a local moonshine made
of maize, in 55-gallon drums, laughing all the while. In the makeshift market, one could buy baskets of dried caterpillars and deep-fried rats broiled on
sticks. And yes, in the background of such scenes, bags of milled maize, the
makings of nshima (or mealie-meal, as it is called in its uncooked state) sat
fat under the sun, even in this dispirited little squatter town.

By the summer of 2003, close to a year after the height of the GM food
aid crisis, the issue should have been moot anyway. Harvests had doubled
over the previous year, bringing in 1.2 million tons of maize, and Zambia
seemed to have weathered yet another stormy dip into the dangerous territory of severe food insecurity. But far across the Atlantic, on the fattened
shores of a United States still suffering casualties in Iraq, President Bush
found it still enough of an issue to close a commencement day speech with
the topic.

The argument that this was a technology that could help rather than
hurt Africa was also a bit misleading, at least according to those on the
ground there. As the agriculture professors at the University of Zambia had
revealed, the Bt trait that killed off caterpillars in the fields might actually
reduce a foodstuff that Zambians regularly ate. And regardless of what kind
of labor or chemical-saving compound could be engineered into GM seeds,

they would never manage to even sprout in the conditions observed during the droughts in some Zambian fields.

While images of American industrial agriculture abound with lush stands of endless grain, what feeds that lushness behind the scenes is not just expensive inputs like chemical fertilizers, pesticides, and high-tech engineered seeds, but a biological basic that along with light from the sun is necessary for any plant to grow and thrive: water. In the developed world, agricultural producing regions rely upon an expansive (and expensive) system of snaking irrigation technology, designed to precisely deliver water to plants when and where they need it.

Certainly, droughts sweeping the American Midwest have had devastating effects and triggered crop failures, but in places like Zambia this phenomenon is taken even further to the extreme. There, most farmers toil without any system of irrigation in place, so blessed has been the climate and reliable have been the rains in recent history. Indeed, many farmers in Zambia merely wait for the rains to begin, then walk out to their fields, poke a hole in the ground with a finger, and drop in a kernel of last year's corn (or if lucky, some locally produced hybrid seed), and watch for it to sprout. But now, as global climate change sets in and droughts and floods replace the steady rain from the skies, tender sprouts can get caught in weeks-long desert conditions, in which the soil surrounding them hardens like a vice, choking out the start of the season's crops.

Food Is Not Enough to Stop Famine in Its Tracks

That is one of the reasons that Mundia Sikatana and other expert observers from within the country decry the need for GM seeds. A high-tech band-aid will never be able to heal the low-tech wound inflicted by decades of isolation from economic prosperity—and indeed, dependence, whether justified or not—on foreign food aid, not to mention the perils of wildly fluctuating weather conditions born of global climate change.

Thus, before GM seeds could have any impact on agricultural productivity in Zambia, basic conditions, such as the provision of irrigation systems, would need to be met. There is much name-dropping when it comes to the hope of drought-tolerant GM crops, but qualities like resistance to dry spells—which should be distinguished from all-out lack of water—are difficult to engineer into food plants and take a toll on their productivity as well.

Because traits such as drought tolerance that would benefit the developing world are often superfluous to the cash-cow industrial farming needs of developed nations, the financial incentives needed to carry such products down the long road of research, development, and regulatory approval have thus far been invisible as far as the eye can see. Certainly, great effort has been made to discover such applications, and likely quite useful products languish on laboratory shelves awaiting their "angel donor" or a truly kinder, gentler time when resources are redirected toward public ownership and investment in biotechnology. And if and when industrialized nations do recognize a need for such traits in the face of global climate change, such important commodities will likely become locked up behind the prohibitive patent regulations that prevent poorer nations from taking part.

Regardless, even a return to the open-source seed scenario present in agronomist Norman Borlaug's Green Revolution and a sudden treasure trove of GM traits benefiting third world farmers is unlikely to stop the scourge of world hunger in its tracks. Even in wealthy nations, stomachs rumble and citizens go to sleep hungry, although surplus food is bountiful. The root causes of hunger extend far beyond the provision of adequate food supplies. As the Indian Nobel laureate Amartya Sen—who received the prize in 1998 for his longstanding contributions to economics and social welfare—argued in his acceptance speech, "The dynamics of income earning and of purchasing power may indeed be the most important component of a famine investigation. This approach, in which the study of causal influences on the determination of the respective incomes of different groups plays a central part, contrasts with an exclusive focus on agricultural production and food supply, which is often found in the literature on this subject."

Promoters of biotech will argue that that is precisely where their focus lies; that the issue of gross yield of foodstuffs aside, it is really the increased income that GM seeds can bring to the small farmers that will allow them to finally prosper in the world of twenty-first–century capitalism. Not without a smug nod in his direction, it is frequently Sen's own native India—a new entrant into the global marketplace and one of the top five growers of GM seed—that is often put forth as a poster child in this effort. Even in a nation wary of GM food such as India, biotech traits can still solve the riddle of world hunger, industry leaders say. But once again, it is what is on, and indeed in, the ground that counts. While India may be several strides ahead of African nations in its economic development, the messy entanglement of the GM debate lies no less settled there.

"Our Seeds Know Our Soil"

Knowledge, Power, and Resistance to GM Food in India

In contrast to Zambia, where poverty predominates like a heavy blanket over the landscape, India is a country of contradictions. Roughshod rickshaw drivers chatter away on sparkling new cellphones as they brazenly dodge cows, bicycles, and the odd elephant that share the streets with belching buses and shiny late-model cars. In a country that has outsourced its information technology workforce and insourced work that much of the developed world neither wants nor can afford to do, the issue of GM food spans as diverse and colorful a spectrum as everything else.

Which is not to say that it necessarily registers high on the radar screen of the average Indian citizen. In Zambia at the time of the food aid crisis, even taxi drivers were willing and able to weigh in on the issue of GM food. In contrast, in India the discourse surrounding GM food festers in isolated pockets around the country, where dueling friends and foes of the technology build isolated empires that often register greater impact outside their own country than within.

Leading the opposition to GM foods in India are a triumvirate of Western-educated scholars who have taken what the West has to offer (mostly degrees and a certain level of status) and returned to their native country intending to use their knowledge and power to defend, or at least define, a better direction for their country than following in the footsteps of the biotech multinationals. These are the voices that often speak of the indigenous food and farming traditions of their diverse nation. They speak of the potential for India to "go it on its own," keeping its knowledge intact rather than selling it out to the sharks of corporate science, which would rather steal than share the wealth of information that the country has amassed over its generations. Despite frequently speaking on the part of the poor and impoverished, the leaders of India's anti-biotech bastion remain for the most part members of the elite, living comfortably and commanding a position from which their voices are heard round the world.

Vandana Shiva, Fiery Foe of Biotech Food

The best known of these leaders is Vandana Shiva, the fierce, full-formed figurehead of the global anti-biotech crusade. Shiva, a quantum physicist who received her PhD from the University of Western Ontario in Canada, stands at the helm of an organization known as the Research Foundation for Science, Technology and Ecology, which has its offices scattered throughout a warren of flats in Hauz Khas, a trendy yet gritty section of south New Delhi. Most people know Shiva more for her outspoken and impassioned public lectures that rail against multinational capitalism and in defense of indigenous knowledge than for the smaller scale work of her organization on the ground in India.

Shiva rose to fame with her involvement in the Chipko movement, back when "tree hugging" wasn't an epithet but a political strategy on the part of poor women who were trying to save the forest for food and fodder in the face of encroaching mining and dam development in rural India. Shiva's first notable book, *Staying Alive: Women, Ecology and Development*, published in 1989, drew connections between environmental destruction and the social oppression of women and spearheaded a movement of environmental feminism known as ecofeminism.

In its watered-down version—a criticism often flung at Shiva—ecofeminism is interpreted as overly embracing the woman-nature connection. Such a connection might fit within the context of indigenous frameworks, but can seem alternately too touch-feely or overly oppressive in the United States, a nation fueled by career women and a "nature as natural resource" mentality. From her involvement in Chipko and work with women farmers, Shiva easily moved to the helm of the anti-biotech movement, preparing that ground by giving voice to what she called, in a book of the same title, *The Violence of the Green Revolution*, confirming disturbing trends noted by others.

Saving Seeds and Promoting Sustainable Alternatives

During the 1990s, as GM food advocates gathered steam and the biotech industry prepared for a full-fledged launch of their patented products, Shiva was busy writing about the threat of biotechnology to India's biodiversity. She was also busy stockpiling a seed storehouse on her educational farm, called Navdanya in Uttarakhand, outside her home city of Dehra Dun. The farm serves as a resource to small farmers and seed savers, and helps to promote Shiva's mission through workshops that draw international participation. In neatly stacked tin flasks, more than 300 rice varieties and a smattering of seeds of wheat, barley, millet, and oilseed crops sit on the shelves of a small outbuilding surrounded by fertile fields supplying the farm and testing out new varieties of seed. A small forest backs the farm, and a mango grove boasting nine different varieties of the succulent fruit faces the front.

Navdanya, which means "nine seeds" in Hindi, is much more than just a seed bank. Community groups gather there to learn sustainable farming techniques. More than 70,000 small farmers are primary members of the organization, committing to send in seeds in subsequent years or share their seeds with at least two other farmers following their receipt. It is this approach, far more than the quick fixes of biotech, that Shiva believes will improve the lives of India's 115 million farmers, a livelihood that supports 70 percent of the country's population, where the average income hovers around $900 per year and 850 million citizens live on less than $2 per day.

Shiva's statements on biotechnology and agriculture have a heavy bent against industry, with titles like "Food Democracy v. Food Dictatorship" and "Biotech Wars: Food Freedom v. Food Slavery," in which she claims that "the spread of GM foods reflects the spread of food dictatorship, not free and informed food choices. Contrary to claims made by proponents, GM foods are undermining food and nutritional security, food sovereignty, and food democracy."

Shiva's Attack on Golden Rice

Though the biotechnology industry would like Shiva to disappear—to take her sari'd self back to her native northern India hills and return to hugging trees—because of her stature as a scientist and her international following, media outlets have picked up on her message. "GM rice promoters 'have gone too far'" proclaimed a headline in the United Kingdom's Guardian in February 2001, following the announcement of Ingo Potrykus and Peter Beyer's successful creation of Golden Rice. "Claims by the biotech industry and some U.S. politicians that genetically engineered 'golden rice' would save the sight of 500,000 children a year are exaggerated, according to the Rockefeller Foundation, which is funding the rice's development," wrote Paul Brown, the Guardian's environment correspondent, in a special report on the GM debate.

That claim was, in some sense, Shiva's doing. Shortly after the successful development of Golden Rice was announced, Shiva had set to work debunking its "biotech band-aid" approach to nutrient deficiency in India. "Unfortunately, Vitamin A rice is a hoax, and will bring further dispute to plant genetic engineering where public relations exercises seem to have

replaced science in promotion of untested, unproven, and unnecessary technology . . . a far more efficient route to removing vitamin A deficiency is
biodiversity conservation and propagation of naturally vitamin A rich plants
in agriculture and diet," Shiva wrote in a position paper titled "The Golden
Rice Hoax: When Public Relations Replaces Science."

Included was a scientific analysis showing that some commonly used
Indian ingredients—including coriander, amaranth, carrot, pumpkin,
mango, and jackfruit—could provide sufficient vitamin A dosage if encouraged in the diet, and that "in order to meet the full needs of 750 micrograms
of vitamin A from rice, an adult would have to consume 2 kg 272 g of rice
per day. This implies that one family member would consume the entire
family ration of 10 kg from the PDS [public distribution system] in 4 days to
meet vitamin A needs through Golden Rice." Besides, she argued, even if
Golden Rice did put a dent in vitamin A deficiency, further entrenching
dietary dependence on rice alone was sure to lead to additional micronutrient deficiencies that might be avoided by supporting a more varied diet rich
in fresh fruits and vegetables, especially leafy greens that were easily cultivated by women and amenable to small gardens.

Peter Melchett, then executive director of Greenpeace U.K., had directed Shiva's dispute to Gordon Conway, an applied ecologist with a deep
development history who was then at the helm of the Rockefeller Foundation, the organization that had partially funded the creation of Golden Rice.
Conway read Shiva's complaints, and in a letter to Greenpeace dated January
22, 2001, attempted to dispute much of her attack. Bioavailability of vitamin
A in leafy greens can often be low, he argued, and for poor families, such
foods might be scarce anyway, especially in times of drought. Animal
sources of vitamin A were ruled out for vegetarian households, which are
common across India. Low levels of B-carotene in Golden Rice were based
on first-generation technology, and as of his writing, new and improved
strains of Golden Rice had been developed, shifting Shiva's calculations on
the intake of rice that would be needed.

Shiva's Critique Gathers Momentum

But despite all of this, there was one area in which Sir Conway (he was knighted by Queen Elizabeth II in 2005 for his service to agriculture and development) was willing to concede. "Finally, I agree with Dr. Shiva that the public relations uses of Golden Rice have gone too far," he wrote. "The industry's advertisements and the media in general seem to forget that it is a research product that needs considerable further development before it will be available to farmers and consumers. I hope these comments and the enclosures are useful."

Useful they were, as Greenpeace broadcast far and wide Conway's concession that Golden Rice had "gone too far." It was via this path that Conway's quote found its way to the headline of Paul Brown's story in the *Guardian*. Another case of liberal media bias, grumbled biotech supporters, but it was seemingly too late: Shiva's genie was out of the bottle. The following year, *Time* magazine would feature her and her message in the "Heroes" section of a special "Green Century" issue, coinciding with the World Summit on Sustainable Development in Johannesburg, South Africa. Admittedly, *Time* toned down Shiva's anti-GM message, saying "Her pet issue these days is preservation of agricultural diversity. It is under assault, she says, from global companies that encourage farmers to grow so-called high-yielding crops that result in a dangerous dependence on bioengineered seeds, chemical fertilizers, and toxic pesticides."

But biotech advocates remained rankled. A report from the conservative CNS News service in response to the *Time* coverage claimed that the magazine's "decision to honor an Indian organic farming activist for her opposition to genetically modified foods is drawing fire from critics around the world who accuse Vandana Shiva of advocating a return to the 'days when people died like flies.' " No matter that the primary critic they referred to was Barun Mitra, a stand-in man for Monsanto. (Mitra was also the head of a libertarian think tank in Delhi, and a contributor to a book entitled

Global Warming and Other Eco-Myths.) With Shiva's robust and proudly Indian profile in *Time* magazine following on the heels of Ingo Potrykus's face gracing its cover, it was clear that her message was there to stay.

Suman Sahai, Scientist Turned Activist

Not far from Shiva's south Delhi headquarters in Hauz Khas, a lesser known yet similarly powerful anti-GM scientist holds court. The contrast between the medieval chaos of Shiva's neighborhood and the upscale, suburb-like luxury of Suman Sahai's post in the Sainik Farm settlement reflects the difference in approach that these two similarly schooled and like-minded women take to the issue of GM food. Shiva is a recipient of the Right Livelihood Award, the so-called "alternative" Nobel Prize, for "placing women and ecology at the heart of modern development discourse." Sahai was named a Norman Borlaug Award winner for her work "promoting scientific awareness relating to biotechnology." Shiva holds court with Prince Charles, a fellow advocate of organic farming and staunch opponent of GM food. Sahai's organization is chaired by M. S. Swaminathan, the legendary "Father of the Green Revolution" in India.

Despite these differences, these two women share far more in common with respect to disdain for GM technology than what holds them apart, and despite her more conventional nature, Sahai's anti-GM message is no less powerful. Sahai, a geneticist who was educated in India but left for some years to become a professor at the University of Heidelberg in Germany, stands at the helm of Gene Campaign, an organization whose mission is "to work for the empowerment of rural and *adivasi* [indigenous tribal] communities; and for laws and policies supportive of food and livelihood security." In addition to its representation of poor and small farming constituents in policy-making, Gene Campaign also seeks to "raise a public debate on the new biologies and their application, and to formulate policy in this area."

TRIPping Over Property Rights for Plants

More low-key than Shiva's flamboyant demeanor but no less dogged in her approach, Sahai, through Gene Campaign, has wielded a ferocious battle against the intrusion of GM technology into her native land. The first fight she took on arose within the context of intellectual property rights for Indian farmers. In the wake of the World Trade Organization's TRIPS (Trade-Related Aspects of Intellectual Property Rights) agreement, which India had signed onto in 1994, the country was bound to adopt a procedure for the protection of new plant varieties, either by issuing patents, by developing an effective sui generis (Latin for "of its own kind") policy of its own, or through some combination of the two.

Lurking in the background of TRIPS was the Union for the Protection of New Varieties of Plants (UPOV), a consortium of wealthy industrialized countries that had banded together to issue commercial breeders ownership over "new" varieties of plants. While its mission statement claims benevolent goals—"to provide and promote an effective system of plant variety protection, with the aim of encouraging the development of new varieties of plants, for the benefit of society"—UPOV's Internet home page is in fact quite blunt about its intention: "the objective of the Convention is the protection of new varieties of plants by an intellectual property right."

Obviously, in a country such as India—where small-scale farmers predominate, and many, if not most, use and sell seeds they have bred themselves—adopting a system in which large-scale seed companies retain intellectual property rights to seeds would dramatically impact the livelihoods of a large segment of the population. Sahai recognized this immediately and, like Shiva, mounted a vocal campaign against patenting and privatization in India's seed industry.

After seven years of diligent political and educational campaigns, Sahai saw victory on the horizon. The Indian government had opted for the sui generis option under TRIPS, and with intensive urging from Gene Campaign, had developed its own regulatory policy, embodied in the Protection of Plant Varieties and Farmers' Rights Act of 2001. The latter clause of the act's title carried the greatest weight. According to Sahai, "The Indian law was the only law in the world that provided for comprehensive farmers' rights including the rights of farmers over seeds, as well as protection against bad seeds provided by breeders and the right to compensation."

According to the text of the law, farmers in India have the right to "save, use, sow, resow, exchange, share or sell farm produce including seed of a variety protected under this Act." Conversely, through the act, farmers are also protected from false claims or shoddy performance of a seed variety that they purchase. The so-called terminator technology—a genetic modification causing subsequent seed to remain sterile, which had been developed by the biotechnology industry to further guarantee against seed saving by farmers—was explicitly forbidden. In the case where a new seed variety was derived from a previous farmer's strain, the breeder would be required to pay a royalty into the National Gene Fund, a form of benefit sharing recycled back into farmers' welfare issues, such as maintenance of gene banks or compensation for crop failures.

This type of legislation was completely at odds with the UPOV model, which gave rights to breeders only, not to farmers, and relied on patents rather than sui generis protections. Gene Campaign's first challenge had ended a success, or so it seemed. But there was to be far more work for Sahai in coming months. No sooner had the Protection of Plant Varieties and Farmer's Protection Act of 2001 been passed than the Indian cabinet made an about-face. Circumventing parliamentary discussion, the cabinet approved an intention to join UPOV, a decision that would necessitate overriding the farmers' rights guaranteed by the protection act.

Sahai was dumbfounded, and Gene Campaign was not about to let go of the fight. Seven years of struggle, of bringing different parties to the table, and of pushing to craft ground-breaking legislation that would protect poor and vulnerable farmers was about to be shelved by an abrupt cabinet decision made without parliamentary approval. Gene Campaign made phone calls, pleaded, and appealed with the Ministry of Agriculture to undo this turn of events, but to no avail. Determined not to stand by in the face of this unethical and misguided course of action, in October 2002—just over a year after the passage of the protection act—Gene Campaign filed a Public Interest Litigation petition in the Delhi High Court, claiming that the decision to join UPOV was both illegal and unconstitutional. The government responded that it fully intended to follow through with UPOV membership, albeit under the less-restrictive 1978 convention, rather than the more recent 1991 convention. But UPOV is not an easy club to join, and as of its most recent member listing in late 2007, India had still not been admitted into its ranks.

Bringing the Stakeholders to the Table

Fighting UPOV was only the first of Gene Campaign's many legal and political battles in the fight over GM in India. But Sahai's mission with Gene Campaign was not merely antagonistic. Using her background and experience as a scientist, and driven by her commitment to the poor and hungry in her country, she steered the organization's operation in the direction of finding constructive solutions and toward stakeholder participation in policy-making.

In this regard, in the fall of 2003, on the tenth anniversary of the founding of Gene Campaign, Sahai organized a two-day national symposium on "The Relevance of GM Technology to Indian Agriculture and Food Security." In attendance were friends and foes, rich and poor, people whose interests in biotechnology and indeed even in agriculture varied widely. From scientists and lawyers to members of parliament and poor farmers, from foreign ambassadors and seed industry executives to grass-roots environmentalists and organic farming associations, they attended the symposium, whose goal was to provide an opportunity for all views to be freely expressed. M. S. Swaminathan, the legendary Green Revolution leader, gave the inaugural address, and the president of India's National Academy of Agricultural Sciences made a keynote presentation. A summary of recommendations on how biotechnology in India should proceed was produced, focusing on the need for increased internal regulation and a shift in research priorities to projects of national value. It seemed that Gene Campaign had succeeded with its intended goal of bringing different stakeholders together to speak to one another, if not see eye to eye.

Troubles with Bt Cotton in India

The success of the anniversary conference was in sharp contrast to another event that Sahai had organized through Gene Campaign several months earlier. At that meeting, held at the India International Centre in Delhi, scientists, non-governmental organizations (NGOs), and representatives from both the Indian government's Genetic Engineering Approval Committee (GEAC) and Monsanto and its Indian subsidiary, Mahyco, had been invited to discuss the outcome of India's first GM cotton harvest.

A year earlier, after lengthy procrastination and a lawsuit filed by Gene Campaign charging a lack of adequate government oversight, the Indian government had finally approved Bt cotton for commercial use in the country, beginning with the cotton season of 2002. In some senses, this was merely a formality. While the government had held off on official approval pending years of field trials, Bt cotton had been alive and well in India since 1995, when the Indian government had first permitted the Maharashtra Hybrid Seed Company (later to become Mahyco-Monsanto Biotech Ltd.) to import 100 grams of an early Bt strain of cotton seed from the United States. By 1998, Mahyco and Monsanto had been formally intertwined, and small-scale field trials of Bt cotton had been authorized by the government, despite public protest. Two years later, forty large-scale test sites in six Indian states were under way, and Mahyco was pressing for commercial release in the face of growing protest from Greenpeace. An additional year of field testing was authorized by the government, but by then it was too late.

In October 2001, during the final large-scale field trials taking place prior to commercial approval, Mahyco discovered that Bt cotton was already in widespread commercial use in the state of Gujarat, and that seeds possessing the proprietary Bt trait were being sold by another major seed distributor, Navbharat Seeds Pvt Ltd. It was not just the current season that would be affected by this transgression but at least several subsequent years as well, since the cotton harvested from the initial batch of renegade seeds would yield its own bounty of seeds in subsequent generations.

Making matters worse, these "black market" Bt seeds were reported to be selling for even *less* than the price of normal hybrid seeds (because they were unregulated and thus an underground product), and of course far, far less than Mahyco and its big brother Monsanto had planned to charge upon approval. Demanding an official response to this dramatic infringement, the GEAC ordered that the illegal cotton fields be burned. The ruling was never enforced, however, and the cotton was left to stand and go to seed. By the end of the winter, it would be a moot point anyway, as the GEAC would by then have approved Bt cotton for commercial use beginning in 2002, and Gene Campaign would be pressing ahead with another lawsuit charging the government with inadequate regulatory and safety precautions.

And so, perhaps, it might not be surprising within this context that attendance was spotty at the first Gene Campaign–sponsored event in 2003, which was intended to discuss the post-approval status of Bt cotton in India and to present the results of a survey that the group had conducted on the performance of Bt cotton in the states of Maharashtra and Andhra Pradesh. Although invited, the GEAC chair declined to attend, and representatives of Mahyco-Monsanto were visibly absent, at least in official form.

The Tables Turn and Monsanto Sends in Their Naysayers

The meeting went on as planned, at least for a few minutes. Then, shortly after the opening discussion began, a parade of unruly farmers entered the room, mumbling in Telugu, the official tongue of Andhra Pradesh. One farmer—seemingly the appointed leader—began to pick a fight with the convener over the validity of the study. The media captured the chaos, as the intruders gushed enthusiastically in front of the cameras about their positive results with Bt cotton and how eager they were to plant it again.

This, of course, contradicted the results that Sahai had been poised to present, based on a carefully conducted survey of a random sampling of one hundred farmers using the technology for the first time in two of the country's major cotton-growing regions. Her results showed that both yield and quality were compromised in the Bt-engineered strains and that the plants were plagued by premature boll drop. In addition, her results showed that the plants were also plagued by infestations of the pink bollworm pest, which were apparently resistant to the strain of Bt encoded in the cotton; the plants had been engineered to attack the gut of the green bollworm instead. (Later scientific studies indicated that the pink bollworm possesses three unique mutations making it immune to the effects of Bt.) When all was said and done and the bills for seed and pesticide had been paid, the Bt cotton farmers in Sahai's study ended up faring far worse than those who had stuck with the conventional hybrid strains, losing an average of 79 rupees per acre in return for their experiment with the new seeds.

If Bt cotton had performed so dismally, as Sahai's study had shown, then why did these apparently enthusiastic farmers suddenly descend out of nowhere on the meeting? Had Sahai's random study approach not been so random after all? That was precisely the message that these farmers were struggling to get across to the media, which was all too eager to capture on camera this scene of professional chaos and conflict.

As it turned out, it was a message that the farmers had been paid to send. When asked, some of the farmers confessed that they had arrived from Andhra Pradesh on an all-expenses-paid train trip sponsored by Monsanto. They admitted that they had been coached to present a rosy view of Bt cotton and to disrupt the meeting, speaking in favor of Bt cotton's continued use. Once their cover had been revealed and the stage cleared, the farmers filed out. An air-conditioned bus was waiting for them, along with their escort, Barun Mitra—the Monsanto spokesman and critic of Vandana Shiva.

But Sahai was not deterred. Rather than step back from the spotlight, she planned a bigger and better meeting the following year to celebrate her organization's decennial anniversary and to probe further into the future fate of agricultural biotechnology in India. This time, all the stakeholders showed up at the table. Shortly after the meeting's success, she received the Norman Borlaug Award, bestowed in honor of the Green Revolution hero by Coromandel Fertilisers, India's second largest phosphate manufacturer and a supplier of chemical fertilizer and pesticide products. Monsanto's feeble attempt to stop Sahai had obviously failed.

Devinder Sharma, Defender of India's Food Security

In contrast to Sahai and Shiva, the third pillar of India's anti-GM elite is held by Devinder Sharma, a plant geneticist by training who early in his career stumbled into the position of agricultural editor for India's liberal but highly respected *Indian Express* newspaper, a known training post for independent and critically minded journalists. After a decade of work unearthing the skeletons of food and agricultural politics and shedding light on the impending impacts of globalization in India, Sharma left the newspaper to consolidate his ideas into a book entitled *GATT and India: The Politics of Agriculture* (GATT is the General Agreement on Tariffs and Trade). He intended to return to his journalism post upon the book's completion. He never did.

Instead, driven by his conviction that food security in India was under siege by the forces of globalization and biased trade policies pushed by developed countries, Sharma founded his own research and policy think tank, the Forum for Biotechnology and Food Security. He continued his research and reissued his book in a second edition, *GATT to WTO: Seeds of Despair*, with a ramped-up warning message about the dangers of global trade policies driven by dominant economies and corporations.

Like Shiva and Sahai, Sharma takes issue with the role played by intellectual property rights—bred into GM food technology by the companies and countries that proffer these products—in the context of hunger and poverty. Sharma has not always been stridently anti-GM, however. His background as an agricultural geneticist allowed him to initially see positives within the science. It was only upon analyzing how agricultural biotechnology has unraveled and how its implementation has played into the dominance and control of the world's major staple crops by a few countries and corporations that he began his crusade against GM technology.

GM, Globalization, and Food Entitlement for the Poor

Sharma's view of GM is that it is but one weapon in the arsenal of globalization that is beating developing countries into food insecurity and solidifying the bloc of hungry bellies in the southern half of the world. As he stated in an interview with *In Motion Magazine* in 2003 on the subject of the World Trade Organization and agriculture, Sharma believes that "the basic idea, or the basic focus, today, is to keep one half of the world hungry, because you can only exploit the hungry stomach."

This may sound like a radical and outlandish prospect, that trade policies and global superpowers have as their agenda the starvation of large swathes of the world. Indeed, it flies in the face of many multinational corporate mission statements, such as Monsanto's, which promises "We apply innovation and technology to help farmers around the world be successful, produce healthier foods, better animal feeds and more fiber, while also reducing agriculture's impact on our environment." It also flies in the face of the vast resources dedicated by the U.S. government to famine-watch programs like FEWS NET (the Famine Early Warning Systems Network). Indeed, as the fray in Africa revealed, food aid is generously given, whether

GM or not, in an abundant flow from north to south. What, then, could Sharma possibly be referring to?

To support his accusation, Sharma draws on the work of his country-man and Nobel laureate Amartya Sen, propagator of the "theory of entitle-ment" that altered the course of thinking on the economic analysis of hunger. Arguing that hunger arises when an individual's food entitlement falls too low, not necessarily when food supplies fail, Sen showed that brutal famine can exist side by side with ample and abundant stores of food. Food entitlement, in Sen's conception, refers not to a person's moral or ethical right to be fed but instead to his or her ability to procure food using his/her "endowments" such as assets or labor. In the food entitlement model, food can be procured through one or more of the following pathways: growing one's food, working for one's food, buying food, or being given food. When food access via any combination of these four channels does not meet an individual's nutritional needs, hunger results. Sen's model was revolution-ary because it contradicted the Malthusian condemnation "too many people, too little food," which had colored much of food aid policy for the third world and indeed, bled through into the promotion of GM food.

Sen's model of food entitlement was cogent to Sharma's accusations on two counts. First, it suggested that producing more food might not be the magic bullet that would solve hunger in India, an argument that Sen made and Sharma applied in the context of chastising the significant subsid-ies provided to rich Western farmers that glutted the market and brought domestic prices down in India. Second, it emphasized an individual's need for food self-sufficiency, in that food entitlement was decoupled from the amount of food available but tightly linked to an individual's personal con-trol over obtaining it.

GM food would tip the balance in the wrong direction on both ac-counts, according to Sharma. If the companies' claims to increased produc-tion stemming from GM technology were true, it would pump even more food into an already saturated system. (Sharma cited statistics on India's simultaneous surplus of more than 50 million tons of wheat and rice, while 320 million people inside the country starved.) This would only serve to further reduce the price poor farmers would be paid for their harvests. And given the types of GM crops developed and available, coupled with the stringent intellectual property regimes protecting their profit for their producers, Sharma was convinced that the GM project would do nothing to alleviate hunger in India and indeed would only amplify the gap between hunger and food security that so many of his fellow citizens struggled with.

From Green Revolution to Suicide Epidemic

As a portent of the potential harms that might come from embracing GM food, Sharma points to the link between the presumed success of the Green Revolution and the wave of suicide epidemics taking place on the country's farms. While Sharma concedes that the Green Revolution had been instrumental in spinning India out of a cycle of food insecurity in the 1960s and 1970s, he harbors no illusion that a similar tack using GM technology would bring positive results. As Shiva and Sahai also point out, the Green Revolution left in its wake a wave of unsustainable chemical dependence. As these chemicals slowly became impotent against the evolutionary power of insect pests and yields plateaued, simultaneously consuming more and more of farmers' economic resources, the very same chemicals began to kill farmers rather than feed them.

Between 1997 and 2005, more than 150,000 farmers in India committed suicide, plagued by unending debt, crop losses, and a lack of insight into how to extract themselves from this downward spiral. Biotech companies point to pests as the problem and GM traits as a solution—but it is really the debt-intensive cycle of input-intensive agriculture that has brought things to the brink of this rash of suicides. After all, the thing that is plaguing farmers is not just crop failures but the intensive burden of debt that they have incurred following experts' advice, embracing new technologies that were meant to solve problems rather than start them.

CHAPTER 5

Feast or Famine?

The Politics of GM Food Aid in Africa

On a frigid January evening in 2003, as a new year overshadowed by the looming threat of war got off to a tumultuous start, CNN's *MoneyLine* newscast began with the ominous lead "Stocks down again amidst uncertainty of possible war." Suzanne Malveaux, CNN's White House correspondent, quickly launched into the heart of the matter: "The White House is engaged in a full-blown public relations blitz, trying to make their case not only to the American people but also to U.S. allies." The report rolled on, dipping back and forth into the politically tricky issue of dwindling support and even outright opposition to war in Iraq from the nation's European allies. "France yesterday hinted it might use its UN Security Council veto against a UN resolution on Iraq. Germany's Gerhard Schroeder said flat out Germany will not vote for a war."

The news switched gears at that point, warning of an Arctic air front paralyzing states in the north, denouncing the high cost of medical malpractice insurance, and debating whether the country should tap into its strategic petroleum reserves. Results from the previous week's viewer poll were announced: 78 percent of respondents thought American culture was declining, while a mere 7 percent attested to its ascent. And then, it was time for

For example, a typical farmer using Bt seeds in Vidarbha, one of the hardest hit regions within India's suicide belt, will receive a loan for $2500, maybe a little bit more if he is lucky. Of that, $1250 will go to purchase the Bt seeds. Six hundred dollars more will pay for fertilizer and $250 for the laborers needed to sow and till the fields. Pesticide sprays will still be required, and the remaining money won't be enough to cover that. Only 20 percent of farmers' land in Vidarbha is irrigated, and Bt seeds will not perform well unless irrigation is provided, at added expense. If things go right, in the best case scenario, a farmer stands to make a profit of around $1200 off such a project, which must go to feed, clothe, and marry off members of his family during the upcoming year. If things go wrong and the crop fails to perform as promised, the debt will go unpaid, underground moneylenders will begin to lurk, and, for many, a swig from a bottle of pesticide starts to look like the only way out.

Seeking Sustainable Alternatives to GM

Sharma contends that added pressure on the system has been exerted by the shift in the Indian agricultural economy from staples to cash crops, forced by globalization and the pressure applied by subsidies in economically powerful nations overflowing with agricultural surplus. Sharma predicts that in the future, if such conditions prevail, the wealthy West will produce the world's staples, such as wheat, rice, and soybeans, at prices that poor countries can ill afford. Yet the production of these staples will nonetheless remain profitable for their providers. Meanwhile, the increasingly poverty-stricken nations will supply cash crops, such as flowers, coffee, and tomatoes, that more prosperous countries use to temper their tastes and decorate their tables.

It is in this regard, Sharma contends, that India must turn its attention inward, rather than eagerly gobbling up Trojan horse solutions such as the GM technologies of the West. Instead, India must focus on its own needs and resources. In his view, "The only viable path toward sustaining the natural resource base to satisfy the demands of the growing population for food and other agricultural commodities lies in enhancing the potential of domestic agriculture. . . . [I]t requires location-specific technologies and production packages that meet the aspiration of the farmers owning less than 2 hectares on average." In India, 2 hectares is not a small lot; for the 557 million farmers in the nation, the average farm size is only 1.4 hectares.

But these farmers are perhaps the lucky ones. Those most likely to be impacted by a flood of GM seeds into India's fields, and some of those who are most vocally opposed to it, are likely to own no land at all. While scientist-activists such as Shiva, Sahai, and Sharma work at the forefront of the political campaign opposing GM food in India, it is poor and mostly landless farmers there, frequently women, who are working literally at the grass-roots level to promote the alternatives Sharma speaks of.

Seeds at the Center of Life

On the outskirts of the already small village of Zaheerabad, in Medak District on the western edge of Andhra Pradesh state, the rustic offices and meeting space of the Deccan Development Society (DDS) sit surrounded by puddles left from the late-summer monsoon. Despite the recent wetness, it has been a dry summer overall, and across the state, farmers' fields lay fallow, their soil surfaces cracked and dry. It is said that in some places in the state, the groundwater level has plummeted by more than 3 meters. In Andhra Pradesh, less than half of farmers' fields are irrigated, making the proper behavior of summer rains essential to crop productivity.

But monsoons are tricky and unpredictable, especially these days. Early in the rainy season, southwesterly winds carry moisture across the region, sustaining crops in the months between June and September. When these winds die down, winds from the northeast take over, dropping more precipitation from October to December. The careful orchestration and balance between these two moisture-laden forces is critical for ensuring the viability of crops, especially those not under irrigation. Imbalances can lead to alternating droughts and floods and can stop an entire planting season in its tracks.

Shoots and roots are not the only victims when the wet winds fail to deliver. As agricultural productivity plummets, the need for agricultural laborers also dries up, causing mass migrations of the poorer population out of the area. Fodder availability for animal feed diminishes, impacting milk production. Clean drinking water is in short supply, and along the road, women and children can be seen walking, water pots on their heads, to fetch clean water from nearby villages.

But when one arrives at the offices of the Deccan Development Society and surveys the surrounding fields, you see that something there is different. In contrast to the cotton fields and conventional farms gripped in a vice of dessication, the fields that are tended by these *Dalit* women—members of India's infamous "untouchable" caste—are producing bountifully. Something is being done differently here.

Seeds as Wages

In the adjacent village of Pashtapur, a wrinkle-faced woman sits sorting seeds. Her home is tiny by any standards, a mud-walled hut with a dirt floor. In place of furniture, large burlap sacks fill the small rooms. The woman brings out bag after bag of these prized possessions, seeds collected from the fields she proudly tends. "Next season's crop is in the home; this season's crop is in the field," she states through an interpreter. More bags are brought forth, and their spectrum of shades and sizes of lentils and pulse seeds are held up for inspection. Other seeds are sealed for storage, sequestered in fiber baskets topped with a firm coat of medicinal neem leaves mixed into mud.

To say that seeds are at the center of the lives of this woman and others like her would be an understatement. This woman (who goes only by her first name, as many of these illiterate women do) shares the vestibule fronting her small hut with a pair of oxen. She is proud of these beasts; ownership symbolizes her success as well as contributing to it in practical means. She was married before the age of thirteen and has worked as a laborer on the land her entire life. As a Hindu, she does not eat beef, but the manure her pair of animals produces is a vital commodity, not just for soil fertility but also for fuel. Around the village, bowl-shaped patties of cow dung dry in the sun. It is her success with seeds, she says, that has allowed her to buy her oxen. Selecting and saving her seeds carefully has ensured productive harvests under conditions where others' fields have failed.

Other women who belong to her DDS *sangham* (or village association) nod their heads in agreement. "Seeds are our wages," they say. Fittingly, together the women have started a seed bank, which has allowed them to share their success. "Even in this worst drought season, I have given my seed to fifteen *sangham* members!" one nose-ringed woman in a bright sari proclaims. Another member describes how she became involved in nursery raising, supplying plants to the *sangham* and tending to her own crop of fifty fruit and nut trees that provide fuel as well as food.

Many of these women rent rather than own the land they farm, and they have few material possessions. They buy little or nothing from outside their village and live a life of simplicity and sustainability beyond most Westerners' imagination—even, for example, producing their own cooking oil from mustards and other seeds. Generally, they say, they are happy with the land. "Only the land can sustain life, give life. Thus, one must prioritize the land," sums up one member. The circle of women seated at the *sangham* meeting concur and launch into a lengthy description of vermiculture, the worm-composting operations they employ as a supplement to manure. When asked about pest control, a catalog of information issues forth regarding chili-garlic solutions, neem protection, and the power of cow urine and dung.

By working together, these women can hedge their bets and buffer their risks. Seed sharing and a network of knowledge tie the *sangham* together, helping the members get by in rough times. This year, the droughts were so bad that only half of the village got rain, they say, leaving the other half dry. But because they are members of the collective, those who find themselves facing food security can rely on the local, alternative public distribution system, which functions as a community grain fund and promotes regional self-reliance on food, rather than the individualistic tack taken by multinational food aid programs.

The topic of food security leads back to seeds, which by necessity lie at the center of these women's lives. When asked about genetically modified seed, loud murmurs of dissent gush forth in Telugu, the local language. "Our seeds know our soil!" one woman loudly proclaims. "We have no interest in GM crops. These seeds know nothing of our land." Although she may not read and write, genetically speaking, she has the science right. Because the *sangham* has selected generation after generation of each seed variety grown locally on its lands, the information encoded in each seed variety's DNA has been carefully selected to match the growing conditions and needs of the women who farm here. The cycle of cost and inputs imbedded in the GM paradigm are perplexing to these women as well. "If we lose our crop with our own seed, it is not really lost. Imagine if you buy the seed and then lose the crop . . . " one woman breaks off, and the Telugu word for "debt" is murmured round the room, as heads nod.

Spreading the Message through the Media

This scenario may seem like a provincial, Pollyannaish case in point, with little large-scale relevance to the larger, commercial farmers in India who aspire to participate in the world economy. But these women are not Luddites: Indeed, they manage their *sangham* network and seed banks via shiny cellphones and have sought to project their message and their success via the media. While they may not read and write, these women have been taught video and radio skills through a grant-funded initiative of the DDS designed to overcome the barriers raised by their illiteracy. Since 2001, the women of the DDS *sanghams* have run the collective DDS Community Media Trust and have produced more than 500 hours of radio programming on farming as well as five high-quality video productions.

One of these video productions, entitled "Why are Warangal Farmers Angry with Bt Cotton?" has been translated into French, Thai, and German as well as Telugu, Hindi, and English. (Warangal is a district in Andhra Pradesh.) In it, local farmers describe, enthusiastically at first, their experience with Bt cotton. But then, as the growing season progresses, they find that it does not grow as well as the local varieties and is besieged with pests, despite its supposed inherent pesticidal properties. And then at harvest, it produces short fibers, rather than the more valuable long fibers, so it garners a lower price in the market. In the end, Bt cotton fails for these farmers on all accounts—costing them more money for seeds up front, requiring heavy pesticide use despite built-in pesticide claims, and reducing yield and quality.

These findings were quickly countered by a market-research agency study commissioned by Monsanto, which painted a deeply contrasting and rosy picture of the Bt cotton experience in India. And so began a scientific back-and-forth, with various groups presenting numbers wildly at odds with one another, each supporting one side of the Bt cotton claim. And in the meantime, cotton farmers across India continued to commit suicide in alarming numbers.

India's Future with GM Food

Given this backdrop, India did not come easily into the GM fold. Given the country's relatively recent experience with the perils of famine in the twentieth century, along with a tradition that prioritizes small-holder farms that support the livelihoods of two-thirds of the population, India was initially slow to take up GM technology. Lawsuits and public outcry—such as the 1998 "Cremate Monsanto" campaign, in which farmers in the Karnataka cotton-growing region rebelled and burned Monsanto's test plots to ashes— put GM approvals on the slow track in an already circuitous regulatory bureaucracy.

But in recent years, India has approved field trials for many genetically modified foodstuffs, including rice, potato, okra, tomato, groundnut, brinjal (eggplant), cauliflower, cabbage, castor, mustard, and cotton. Though field trials are a far cry from finding these foods on the nation's dinner plates, the willingness of government regulators to welcome GM technology as applied to foodstuffs sends an encouraging signal to the promoters of agricultural biotechnology eager to hasten its embrace in the developing world.

Despite the persistence of its native scientist-critics such as Shiva, Sharma, and Sahai, and the legal battles and lawsuits they put forth, it seems that India—fresh with the energy of its newly won, self-made fortune in the capitalist playing field—will have trouble resisting the lure of GM technology for long. But while high-tech solutions tend to make the news, behind the scenes, India's small-holder farmers, particularly at the lower spectrum of the caste spectrum, continue with their work of finding grass-roots solutions of their own. In the end, this small-scale science may turn out to better fit the needs of the majority of India's agricultural class than the high-tech gene fixes engineered in urban laboratories.

CHAPTER 7

The Scent of GM Papaya

The Rough Road to Biotech Rice

By mid-2003, the United States had filed its World Trade Organization (WTO) suit against Europe, with government officials crowing in the background that Europe's stonewalling was costing American farmers $300 million per year in lost export income. In fact, U.S. corn exports to Europe had fallen dramatically since the wide-scale advent of GM crops from a high of 3.3 million tons in 1995 to a paltry 25,000 tons in 2002. This agricultural bloodletting resulting from Europe's stubbornness, combined with the last-minute backing out of fair-weather friend Egypt in the WTO suit, had led the Bush administration to recognize that it needed to be proactive and start sealing up any gaps with its second tier of trading partners.

Turning to Thailand: The Cost of Free Trade

Thailand is the eighteenth largest trading partner of the United States and had proven its ability to both prosper and follow Washington's suggestions prior to the great Asian financial crisis of 1997. Thus, it seemed as good a place as any to start. Thailand had recently signed an agricultural Free Trade

Agreement (FTA) with China that cut trade tariffs to zero. Weakened and vulnerable as Thailand was in the aftermath of the financial crisis, it seemed it should be relatively easy for the United States to construct a FTA with which Thailand would readily comply. An added bonus was the hope that countering Thailand's recent agreement with China with a trade pact of its own would keep the United States favorably privileged in the face of competition from Thailand's closer Communist neighbor. On October 20, 2003, the Bush administration announced its intent to consult with Congress on the Thailand FTA proposal as mandated by the U.S. Trade Promotion Authority. Then formal pursuit of the FTA could begin, after a required ninety-day waiting period during which additional consultations were to take place.

On the other side of the Pacific, despite the eagerness of the Thai business community to engage in a free trade agreement, even many of those in favor were wary of the potential for it to backfire. "Since the U.S. would like to pry open our markets in agriculture, industries and services, we should try to make sure that we have full access to the U.S. farm market," Pornslip Phatcharintanakul, director of the Board of Trade of Thailand, warned in an article in the Thai newspaper The Nation, not long after pursuit of the pact was announced.

The same article briefly touched on the concerns of Darunee Edwards, a Board of Trade member charged with addressing the touchy subject of genetically modified organisms (GMOs). As it turned out, since 2001, Thailand had enforced a moratorium on commercial planting and trade in GMOs, the relaxing of which would necessarily be a condition of any FTA with the United States. Edwards expressed concern that Thailand thoroughly address, on the basis of scientific evidence, the safety of GMOs before allowing the United States to press them upon the country.

By June 2004, the bilateral free trade talks were scheduled to begin, and the GMO issue had risen to prominence on the agenda, along with child labor laws. Of the two, the rift over labor laws was seen as more easily mended. In the meantime, on the GM food front, Thailand had followed in the

European Union's footsteps and had drafted a labeling law requiring a warning message if any one of any foodstuff's three primary ingredients was composed of more than 5 percent genetically modified material. The United States would surely never stand for this, given the pending WTO suit against the European Union.

As the chief trade negotiator for Thailand, Nitya Pibulsongkram, lifted off in a plane headed for Hawaii—the site of the first of what would become many rounds of free trade talks—demonstrators gathered in front of the Royal Plaza in Bangkok to protest their government's haste in pursuing a free trade agreement with the United States. Present were farmers, politicians, academics, and members of non-governmental organizations (NGOs) who pressed for greater involvement of Thai society, from parliament to farmers and the poor, in deciding the conditions of any trade agreement. An elected official from Bangkok emphasized the need for the Thai government to reach out to its constituents who, after all, had voted these officials into office in the first place. Perhaps the demonstrators needn't have worried. In a matter of weeks, the GMO issue would arrive on Thailand's radar screen in its own right, drawing protests of its own. The free trade talks were destined for a dead end anyway, slated for demise in a political crash landing that two years later would bring the entire Thai government down.

The Telltale Papaya

But for now, the stage was being set as if trade would soon flow freely between Thai and U.S. coffers. As government officials debated the ins and outs of agricultural tariffs, labor laws, and patent protections for pharmaceuticals at their meeting on the Hawaiian beachfront, little did they imagine what was taking place in a Hong Kong laboratory on the other side of the Pacific. There, in the offices of a DNA testing company called GeneScan, fruit taken from a papaya tree growing on a small farm in Khon Kaen province (Thailand's northeastern agricultural region) had undergone testing to detect the presence of genetically modified material. The tests had been ordered by Greenpeace, a persistent foe of genetic engineering around the world.

The results that came back to Greenpeace were positive, despite the moratorium banning the growth of GM plants in the country. By the time the trade negotiators had landed back in Thailand, Greenpeace activists were donning their trademark white safety suits and masks and preparing to raid the research lab in Khon Kaen, where they suspected the transgressive GM papaya seeds had originated. On July 27, 2004, campaigners (as Greenpeace employees are called) scaled the fences at the research station, sealed off its fields, and began to pack the suspect fruits into sealed hazardous waste drums.

It took only two days for Thailand's Department of Agriculture to file formal charges against the Greenpeace activists for trespassing, theft, and destruction of a 20 million bhat (in the range of $500,000) research project. In fact, the project did involve genetically modified papaya, but as a completely confined research project it complied with the government regulation prohibiting commercial use or growth of genetically modified crops.

The research station director insisted that there was no way that genetic material from the research project could have gotten out and contaminated farmers' fields. But the station staff admitted that the station had supplied 2,600 farmers with papaya seeds and transplants the year before. The tree that Greenpeace had sampled was located only 60 km from the research station. And a packet of seeds, the same ones distributed to farmers through the research station, had also been sent for testing by Greenpeace, and these too had returned positive. Somehow, there must have been a mix-up, allowing the GM seeds to errantly end up in farmers' hands.

Vehemently, the director and scientists at the Khon Kaen research station resisted such an accusation, at the same time dismissing the dangers that Greenpeace suggested might come from consumption of GM papaya. The scientists and their families had been consuming GM papaya from the test plots during the experiment and nothing had happened to them, they revealed, echoing the logic that U.S. Trade Representative Robert Zoellick had used in scolding reluctant Zambians the summer before. *We've eaten it and nothing has happened to us*—a scientific reassurance indeed.

Besides, such an argument did nothing to win confidence in the current Thai biosafety procedures for GM research. According to the laws currently in force, human consumption without a permit or planned trial shouldn't be taking place, opponents pointed out, even if it involved only the researchers and their families. Any researchers willing to eat an untested experiment themselves surely could not be trusted to protect the public or to enforce the protocols currently in place.

The threat of an unintended escape of GM papaya from a protected research experiment sent a ripple effect throughout Thailand and soon reached the shores of the European Union. Rumors of whether or not Thai papaya could be trusted to truly be GM-free, as the country's moratorium specified, began to circulate internationally. The government sought to avert what appeared to be a political as well as trade disaster in the making.

The Prime Minister Makes an About Turn

Within a month of the report of GM papaya contamination, Thaksin Shinawatra—the prime minister at the time—took a bold and unprecedented step. Without consultation from parliament—let alone involvement of agricultural policymakers or farmers—on August 20, 2004, Thaksin brashly announced that the days of GM-free Thailand were over. From now on, open field planting and commercial use of GM crops would be welcomed by Thailand—provided, of course, that the crops in question received a favorable rating from the country's Bio Safety Committee. It seemed that the three-year moratorium had come to a sudden and unanticipated end. With the wave of this magic wand, the whole papaya fiasco could now be made to disappear. No rules had been broken, after all, and even if they had, there would be no need to address it anymore.

Of course, the average Thai on the street did not see it that way. With the news announcement of this regulatory U-turn—which had incidentally taken place over a weekend—farmers and activists noisily assembled on the steps of the gold-domed Government House on Phisanulok Road, the meeting place of the Thai cabinet. They were joined there by the leaders of many of Thailand's largest rice export companies, who were worried that the new rules would lead to a quick incursion of genetically modified rice into their paddies, toppling Thailand's foothold as the world's number one rice exporter.

"None of our customers wants to buy GM produce," Wanlop Pichipongsa, one of the export executives present, was quoted as saying in the following day's *Bangkok Post*. "Importers, particularly in European countries, always ask for the GM-free labels or non-GMO certificates for rice and farm products from Thailand," he continued.

Many feared that after the papaya episode, importers would likely be asking for them even more. Relaxing the GM policy in Thailand even further was a move in the wrong direction, most exporters felt. The suspicion that Thaksin's move was motivated by his return from the recent FTA negotiating trip was in the forefront of many people's minds. Why not wait and investigate the papaya incident further, seeing what might be learned from it while reassuring importing countries that Thailand knew how to play by the rules of its own making? Why this rush to push through acceptance of a technology and a regulatory process that had so recently backfired in the country's own fields?

Thailand's Struggle to Prosper— With or Without GMOs

Thais had been stung before. Thais had thought a few years earlier that they were riding the crest of the international development wave, buying into the good life that the West had long promulgated and extended in front of them, only to have it all come crashing down suddenly when economic markets failed. Certainly, Thailand had in many senses made steps out of its rural, agrarian past and into the twenty-first century development model that meant big box shopping malls in the heart of downtown Bangkok, slickly paved roads crossing the country, a brisk luxury tourism trade, and world-class healthcare facilities that imported patients from Europe and the United States for major procedures too expensive at home.

But especially in places like Khon Kaen province, and its capital city of the same name, a large and tenuous rural population teetered on the razor's edge dividing survival from prosperity. At the central market downtown, many of the poor sold goods and wares such as imported cast-off clothes, as well as various versions of the country's famous papaya salad, *som tum*, long a staple of the poor northern regions. These were not the people who stood to benefit from free trade policies, for sure, though nearly all of them de-pended on papaya as a foodstuff and for their nutrition, and many had pa-paya trees growing in their backyards.

Prime Minister Thaksin was full of denials that his hasty reversal of the country's GM policy was motivated in any way by the free trade negotia-tions taking place. His denials came despite the fact that the country's Na-tional Human Rights Commission had obtained a copy of the draft free trade agreement and uncovered one article expressly stating that Thailand would be required to dismantle any restrictions relating to biotechnology products before signing. He was confident in his position, Thaksin claimed, and was awaiting approval of his about-face by the cabinet in its weekly meeting, which would take place the following week. "The government won't let the country miss the biotechnology train," Thaksin confidently told the *Wall Street Journal* at the time, eager to report on the potential opening of a new market for Monsanto and other major purveyors of GMOs. Mean-while, the demonstrators went home, threatening to invoke plebeian forces sufficient to wrest him out of office at the next election if the decision was left standing.

Monsanto's Earlier Experiments in Thailand

Those who sniffed the scent of Monsanto on the trail of Thaksin's decision were on to something. Before the week was up, officials from Monsanto were sent to meet with the Ministry of Agriculture. Their mission was purported to be a clarification of the new policy governing GMOs, but Monsanto was no newcomer to the fields of Thailand. Prior to the commercial GM ban instituted in 2001, Monsanto had field-tested both Bt cotton and corn in the country between 1997 and 2000 under the existing plant quarantine laws that restricted commercial distribution of genetically modified seeds prior to results from field trials. In a foreshadowing of the papaya incident, as Monsanto's field trial of Bt cotton had neared completion in the spring of 2000 under close public scrutiny, an agricultural organization had reported the presence of Bt cotton growing in farmers' fields outside the regulated test plots. Somehow, Bt cotton had either been sold or spread beyond the legal bounds of Monsanto's approved experimental area.

In March 2000, thirty-five farmers' groups working together with Thai NGOs called on the government to institute a ban on all commercial GM crops. The government, hoping to put the matter to rest, granted their wishes. By 2001, Monsanto's permits for test plots in Thailand had been terminated, and the moratorium on commercial GM cultivation had been instituted. No further action was taken on the Bt cotton contamination case and the incident quietly disappeared, until the great papaya escape reignited the country's concerns.

The Contamination Unravels

Even with the stringent containment procedures in place under the moratorium, GM plants and seeds had once again friskily made it out of their meshhouses and beyond the bounds of the field trials. In the Bt cotton incident, those in the know had long suspected Monsanto of maliciously and intentionally violating the ban and distributing the quarantined cotton seeds to open up the commercial market prematurely. But with GM papaya, the field station scientists had no real motive or incentive to prematurely pass out an unapproved product. It had to have been either human error or the work of the wind that had carried pollen from the double-screened greenhouses and outdoor research plots and down the street to farmers' fields.

After several months of research, the evidence pointed to human error. By the end of September, the government's own tests had identified forty cases of GM papaya in farmers' fields, though the final count would end up being far more. By the time all 8,912 samples of papaya tracing their lineage to the Khon Kaen research station had been sampled by the Department of Agriculture, 329 samples from eighty-five farms would be found to contain the forbidden genes.

With the revelation that contamination had not been an isolated incident and that Greenpeace had been right after all, the government moved in to finish off the work that the trespassing activists had begun. The Department of Agriculture outlined an official eradication response, beginning with the destruction of all papaya trees on plantations where the outlawed fruit had been found, and the imposition of a 400-meter quarantine zone around each offending site. All identified cases of contamination would in turn trigger a traceability study in an attempt to identify the original source of the contraband seeds.

At the Khon Kaen research station where the trials had been conducted and the contamination had originated, 1,000 papaya trees were uprooted and buried in a large 2-meter-deep pit. Vilai Prasartsee, the researcher behind the project, looked on in dismay. Into the ground were going the literal fruits of a lifetime of her labor, which—through no fault of her own, she insisted—had erroneously been spread throughout the province and country.

From afar, others wondered why GM papaya was such a big deal. As an export, papaya was but one ingredient in canned fruit cocktail, and a small contributor to each can at that. It was not a staple on the table by any means. Still, export of these tinned fruit-salad products brought in 1 billion bhat each year, and Thailand was positioned as the number one world provider of these products. The top two destinations for Thai papaya exports, Japan and the European Union, were staunchly anti-GM.

Within Thailand, on the other hand, papaya was an important foodstuff. It was the key ingredient of the traditional *som tum* salad—a mix of shredded green papaya, spicy chili peppers, and sliced tomatoes—that was a staple for the poor and the pride of the northeastern regional cuisine from where it originates. There, in contrast to the larger plantations in the rest of the country, papaya was predominantly grown as a backyard staple, a subsistence crop on small family farms, interspersed with many other edible species that would never go to market. In this context, genetically modified papaya seemed less threatening and more of a potential help to poor farmers than a harm to the national economy.

The Battle to Rout Out Ringspot Virus

Only, as was perhaps to be expected, most poor farmers did not want genetically modified papaya. Ironically, their trees were less affected by the plague of papaya ringspot virus sweeping the country than papayas grown on large plantations. Mixed species cropping on small farms reduced the load of insect pests, the pesky aphids that transmitted the virus from papaya to papaya despite the fact that it wasn't even their preferred food crop. If a variety of more tasty fruit trees were planted nearby, as they were on small farms, the aphids had less of a chance of landing on a papaya tree. But in larger papaya monocultures, where there were no alternatives, aphids hopped quickly from one papaya tree to another, hoping to find a better meal while spreading disease rapidly throughout the farm at the same time.

It was this unrelenting infection that genetically modified papaya had been created to cure. Since the mid 1970s, papaya ringspot virus had posed a threat to the trees of Thailand, despite persistent attempts to control it. It showed up first as a mottled mosaic pattern on the leaves of seedlings, later streaking their stems and petioles and scarring the fruit with pocklike rings. But more important, beneath the surface of these symptoms, infection with the virus also stunted the plants, leading to a drop in productivity as they struggled to survive. The only surefire way to beat the virus was to destroy any infected trees at the first sign of attack.

Ringspot Virus Emerges in Hawaii

Thailand was not the first papaya-growing region to be struck by this deadly disease. Papaya ringspot virus had first been identified in the Hawaiian islands in the mid-1940s and had begun to inflict damage on crops there a decade later. In the late 1950s and early 1960s, Hawaiian papaya plantations centered on the island of Oahu were hard hit by the virus, which spread easily from one closely planted tree to the next, devastating yields and threatening the industry. Quickly, papaya farmers moved their operations to the Puna district on the neighboring island of Hawaii, where land was cheap, growing conditions were favorable, and there was no ringspot virus. Freed from the disease, the papaya industry prospered, and by the 1970s, nearly every one of the state's commercial papaya plantations had relocated to Puna.

But there was trouble lurking in the background. As farmers were celebrating their disease-free status down the road in Puna, residential gardeners in the city of Hilo, a mere 19 miles away, were fighting off a nasty outbreak of ringspotted fruit turning up in their backyards. A small government task force was set up and charged with rooting out any offending trees, which were burned to prevent spread of the disease. But the writing was on the wall: The Hawaiian papaya industry could not escape ringspot virus for long. A better cure than burning had to be found.

At about this time, a young agricultural scientist by the name of Dennis Gonsalves, working at Cornell University, was called in by the dean of Agriculture at the University of Hawaii. The ringspot virus was under control at the moment, Gonsalves was told, but the state lacked tools to employ should an outbreak occur. The regular backyard reminders that the virus was alive and active on the island of Hawaii were causing everyone fear. Good hygiene and preventive practices could only go so far, and it seemed merely a matter of time before the mottled leaves and spotted fruit would turn up on Puna's large plantations. Could Gonsalves's laboratory start looking for a cure?

Cross-Protection as a Temporary Cure

Gonsalves took the matter to heart and returned to his research station in Geneva, New York, along with a small $5000 grant to get the project rolling. His first strategy drew on the principle of cross-protection, in which infection with a mild, mutated form of a virus would sometimes confer resistance to the more virulent, damaging strain. This was a common agricultural tactic and had been used to control several viral outbreaks affecting citrus, tobacco, and zucchini. Gonsalves took the pernicious form of the virus found in Hawaii and bombarded it with the mutagen nitrous acid, hoping that in at least a few cases exposure to the toxin would induce just enough change to keep the virus infective but tone down its damaging properties.

After several rounds, he hit success, and two mild strains—papaya ringspot virus HA 5-1 and HA 6-1—were isolated. Tests using these strains on virulently infected trees on Oahu indicated improvement. However, cross-protection was not an effective strategy for use on the island of Hawaii at that point, primarily because the virus had not hit outbreak levels there yet, and even the mild strain caused significant symptoms, particularly during the winter. It was a start, and at least there was now a tool, albeit a relatively impotent one, available in case the virus began to spread.

Achieving cross-protection, although not a perfect cure, was enough of a step in the right direction that it caught the attention of another researcher by the name of Vilai Prasartsee, who was working in Thailand's Northwest Regional Office of Agriculture in Khon Kaen. In 1986, Vilai contacted Gonsalves, hoping for some help. She had been laboring since 1979 to get ringspot virus under control in her country and had found total eradication the only effective method. It was difficult to get farmers to comply with this strategy, however, since infected trees still continued to bear fruit, and many poor farmers balked at the idea of chopping down a productive, albeit stunted and infected, tree. Vilai saw cross-protection as a salvation that might aid her efforts to rid Thailand of this scourge, or at least modulate its damaging effects.

But cross-protection was not successful in the hotter climate of Thailand, and another strategy was sought. Working with Gonsalves's team and funded by grants from the U.S. Agency for International Development (USAID) and the Thai government, Vilai first tried to get farmers to accept a variety of papaya known as "Florida tolerant," which showed some innate resistance to the disease. It grew well in Thailand but was too small and not suited to Thai tastes, and it was soundly rejected. Thais can be picky about their papaya.

Next, the shunned Florida tolerant variety was bred with the locally popular strains, and the team found success. Three strains of ringspot-resistant trees—known as Thapra-1, Thapra-2, and Thapra-3—looked promising. In the end, only Thapra-2 turned out to be suitable for growing in the northeastern region. In 1998, Thailand's Department of Agriculture recommended that it be distributed to farmers as both sprouted seedlings and in seed form. Vilai began the distribution program based at the station in Khon Kaen, never intending that these products would ever cross paths with an experimental, genetically modified variety yet to come.

At the same time that Thapra-2 was making its commercial appearance in Thailand, Gonsalves was releasing the fruits of an even more aggressive strategy to confer resistance to papaya ringspot virus in Hawaii. In the meantime, the feared and predicted outbreak of ringspot virus in Hawaii had come to fruition, nearly wiping out the industry there. Between 1992 and 1998, papaya production on the island of Hawaii fell by half, from 53 million pounds to just 26 million pounds. But Gonsalves had been forward thinking and had been inspired by new developments in the plant biotechnology field taking place around him.

GM Papaya Makes Its Debut

Inspired by work with transgenic tobacco conducted by plant scientist Roger Beachy and at Monsanto, in which transgenic expression of the tobacco mosaic virus coat protein in tobacco plants conferred resistance to viral infection in the crop, Gonsalves had early on decided to take a similar approach with papaya. He had isolated the corresponding coat protein from the papaya ringspot virus. Then, using the tools of biolistics—the original "gene gun" used to shoot foreign DNA into plants cells—his team had inserted the gene for the ringspot coat protein into both the red-fleshed and yellow-fleshed papaya varieties grown in Hawaii. By 1991, Gonsalves had created a single transgenic strain, known as line 55–1. The strain was a transformed red-flesh Sunset variety, in which the imported coat protein gene construct was active, and resistance to infection was high under greenhouse conditions.

His timing was impeccable. In 1992, Gonsalves unveiled his new, GM strain of papaya, engineered to resist ringspot. It appeared to be nothing short of a miracle: As line 55–1 sprouted up in field trials at the Waimanalo Field Station on Oahu, devoid of the devastation inflicted by the virus, the first major outbreak of ringspot was reported on Puna's plantations. However, while results from the field trial on Oahu were promising, Puna's climate was different, and it was uncertain whether resistance would be as

effective there as it had been under the experimental conditions at Waima-
nalo. Moreover, Sunset, the parent strain to line 55–1, was red-fleshed, and
Hawaii's commercial growers preferred the yellow-fleshed strain known as
Kapoho for commercial use.

By 1995, Gonsalves had a transgenic yellow-fleshed variety ready,
which had been constructed by crossing SunUp, one of the genetically modified red-fleshed lines derived from line 55–1, with Kapoho. The resulting GM
strain, known as Rainbow, along with SunUp, its red-fleshed parent, were
planted out in field trials in Puna. Both did remarkably well. Farmers were
suitably impressed with Rainbow, which had the preferred yellow flesh as
well as heavier yields and an earlier ripening time than the original Kapoho.
In 1997, the application process to deregulate transgenic papaya was com-
pleted, and in 1998, Hawaiian farmers began to receive seeds of the transgenic strains to plant on their farms, after attending an education session
and signing a sublicensing form.

Part of the success of GM papaya in Hawaii was the rapid rate at which
farmers adopted it. They were tired of running from one island to the next
trying to keep one step ahead of the virus. Cross-protection had worked only
so far, and the conventionally bred resistant varieties were not those the
market preferred. The farmers were willing to try anything, and for the price
of watching a short educational video on the virtues of GM and signing on
the dotted line, they could go away with the seeds of a ringspot-free future.

A study carried out by Gonsalves's wife, Carol, sixteen months after
the great Rainbow giveaway, indicated that 76 percent of farmers surveyed
had planted Rainbow in their fields. Although her survey was conducted
before most farmers harvested any of the transgenic fruit, in general, farmers indicated willingness to pay money for the seeds next time (they had
been free the first time around) or to try new GM varieties in the future,
should they be developed. These were positive signs. But a large majority of
farmers surveyed also strongly agreed with the need to label their transgenic

fruit so that consumers would be aware that they came from genetically modified trees.

This would never happen, of course, at least not in the United States. As it turned out, GM papaya was not just resistant to ringspot. It also appeared to be theoretically immune from many of the criticisms that were dogging other GM products on the market. For one thing, GM papaya had been developed in the public domain, using freely available funds and driven by university researchers and field stations, not big business. Nobody stood to profit from patent protections imposed on farmers who chose to go GM, at least not in a big way. Second, the genes imported into GM papaya differed from those implanted in the other two heavy biotech hitters on the market, Roundup Ready and Bt. Whereas plants encoding these latter two traits had foreign genes inserted into them that pumped out proteins conferring the requisite properties, the viral resistance encoded in Gonsalves's GM papaya worked at the level of nucleic acid, not protein. Thus, its magic did not involve the presence of a foreign protein, at least not one that conferred its commercial trait.

In fact, when it came to foreign DNA, it was the non-genetically modified papaya that was likely to contain more of it, since when infected, these fruits would be teeming with copies of the entire ringspot viral genome, translated into foreign proteins galore. Transgenic papayas, on the other hand, contained only the viral coat protein gene, an untranslated version at that, which would never make a protein. One of the anti-GM forces' major targets had been diffused: GM papaya appeared poised to be the new Rainbow warrior, mocking the name of Greenpeace's famous floating demonstration ship.

Transgenic Papaya Comes to Thailand

If cross-protection had caught Vilai's eye initially, this new GM strategy triggered a sensation across Asia and beyond. Researchers from countries plagued by the ringspot virus began knocking at the door to Gonsalves's lab, and they were ushered in for the price of bringing along a papaya strain from their home country. Two Thai scientists, Dr. Nonglak Sarindu and Dr. Suchirat Sakuanrungsirikul, were sent under sponsorship of the Thai government's Department of Agriculture. They brought with them the three favored Thai papaya varieties—Khakdam, Khaknuan, and Thapra— and for two years they labored along with Gonsalves to develop GM versions of these strains suitable for Thai tastes.

The work was carried out collaboratively, and in the United States, researchers were coming under greater pressure to sign away the patent rights to their inventions to their employers. Thus, a memorandum of understanding (MOU) was developed between the visiting Thai researchers and the Cornell University Research Foundation, the private arm of Cornell's research laboratories, spelling out who would have claims to what if the research came to commercial fruition. Moreover, while the earlier work with Hawaiian strains of GM papaya had been developed with an eye to future public distribution in the developing world, there were still proprietary techniques used in its making that would need to be secured before the fruits of the research could be released. Like Golden Rice, this was a public project that, unavoidably and inadvertently, had tread in its making upon some highly protected patent territory.

As it turned out, when the Cornell University Research Foundation went looking, they found that a minimum of fifteen proprietary techniques would need to be licensed in order for the strains to receive a "freedom to operate" provision and be made available for public use. Securing those rights would be their end of the bargain; the original papaya strains the researchers had brought from their own countries were theirs. Nobody was looking to profit from the project anyway, beyond restoring their home country's papaya industries to good health.

In 1998, the two Thai researchers who had been sent to Cornell returned with a handful of GM strains to be tested in their home territory. At that time, the moratorium was not yet in place, and anyway, since this was a research project, it would not fall under the later moratorium conditions. Plants were put out first in greenhouses and then in the fields at the Khon Kaen research station. The results were astounding: After decades of watching papaya trees faltering under the burden of viral infection, the transgenic trees stood tall and produced heavily. Nary a spot was to be seen, and in the field, two outstanding lines were selected that were able to stand up to even heavy disease burden.

Targeting GM Papaya in Government Fields

As it turned out, Khon Kaen was not the only place in the country that GM papaya was being grown. Beginning in 1997, three parallel approaches to developing ringspot-resistant papaya using GM technology had been initiated in Thailand. One was Vilai's project, the collaboration with Gonsalves that had been sponsored by the government Department of Agriculture at the Khon Kaen research station. A second project was housed at Kasetsart University, a short distance from Bangkok. There, veteran scientist Supat Attathom oversaw the Plant Genetic Engineering Unit's research field trials of ringspot-resistant papaya strains planted in collaboration with Australian scientists, but funded by a string of public and private interests. The third location was a research unit at the Institute of Molecular Biology and Genetics located at the Salaya Campus of Mahidol University, where suspect samples were later sent by the government for confirmation when Greenpeace first reported the contamination.

But several years later, when activists picked their target, it was Vilai's fields at Khon Kaen that they descended upon, rather than either of the two university locations. By then, Vilai had subjected her GM strains of papaya to nearly all of the requisite tests needed to gain commercial approval in the international context: rat feeding, soil impact, fatty acid profile, vitamin C level, and so on. But that didn't matter. It was here that approved varieties of non-GM papaya seedlings were distributed every spring to local farmers, a scenario that provided the most likely opportunity for just the kind of GM mix-up that Greenpeace suspected could happen.

Besides, Kasetsart and Mahidol Universities would be less politically important targets to take on than the Thai government's own research station. At Kasetsart, Supat had the power and prestige of more than a decade at the helm of BIOTEC, Thailand's National Center for Genetic Engineering and Biotechnology. And while he was a firm supporter of GM papaya, studying and growing it in test plots at his university himself, Supat often spoke of widespread commercialization of GM crops in Thailand with the guarded, noncommittal tone of a seasoned scientist.

He found the GM debates a bit boring, he confessed, with each side clinging stubbornly to its stance, and activists attempting to stir politics into otherwise good science. He went along with intellectual property rights because, he emphasized, it was the usual phenomenon in biotech, whether you liked it or not. In the old days, as a scientist you were educated, you published, and you wanted your material to be well used by other people, he said. But not anymore; since the arrival of biotechnology, patents were the new trend, take it or leave it. At the same time, he emphasized, it was important for countries such as Thailand to be able to use patents to protect what was theirs. Even the NGOs would like to protect the native Thai varieties, he pointed out.

From Papayas to Rice—Thailand's Uncertain Future

Supat wasn't referring to papayas when he spoke of protecting what belonged to Thailand, although patent issues had recently superseded health concerns in the ongoing GM papaya debate in the country. Supat's unspoken reference with regard to NGOs, as well as his own patent concerns, were centered solidly on Thailand's coveted status as the world's number one rice exporter. Just as Vandana Shiva in India had balked at and then successfully fought "biopiracy" attempts by multinational corporations to patent India's indigenous claim to Basmati rice, Thailand held a parallel and even more powerful monopoly on its own fragrant strain of Jasmine rice.

Granted, Thailand was the top supplier of rice to the world, but its berth at the top of rice exporting nations was fragile. Trailing closely behind were Vietnam, India, and the United States. Patenting concerns aside, if approved by the cabinet, Thaksin's proposed GM approval legislation could easily allow GM rice to slip in, threatening to tip the export balance in favor of one of the country's close competitors. It was for that reason, Supat said, that he had slowed down his own research on GM rice, intended to develop strains resistant to drought or salinity. "I don't want to put the country into that kind of situation, meaning that I am a responsible scientist," he said when explaining his decision to abandon GM rice research.

On the heels of its success in alerting the nation to the escape of forbidden fruits of GM papaya, Greenpeace had no doubt recognized this link to rice and utilized it to draw attention to its anti-GM campaign triggered by the papaya outbreak, as national attention to GM papaya started to wane. In February 2005, as lawsuits against two of the Greenpeace campaigners were still pending in the Thai courts but the papaya issue had largely died down, Greenpeace staged a demonstration in downtown Bangkok focused not on papaya but on rice. Children's games, artwork using rice, and rice passports were on the agenda along with a theater play depicting the evils of viruses and other GM taints. A poster of a tomato being traversed by a fish stood on the side of the stage, reminding consumers to "just say no to GMO." Anti-GM scarecrows emblazoned with the Greenpeace logo graced green T-shirts available for purchase at the event. The need to protect Thailand's proprietary rice varieties, not with patents but with forceful legislation banning GMOs, seemed to resonate with many in attendance.

Beyond Thailand, rice holds a key place in world agriculture and food supplies. While wealthier global consumers pay premium prices for imported strains like Thai Jasmine or Indian Basmati rice, poorer people the world over rely on rice for their nutritional needs: One out of every five calories consumed by humans in the world today is supplied by rice. This was what had driven Ingo Potrykus to push for vitamin A fortification in his Golden Rice, but it was also what has made consumer acceptance of such an obviously genetically modified product so sketchy. Even staunchly pro-GM scientists such as Supat see the potential dangers of allowing GM rice to run amok. Without proper regulatory structures in place and key importers willing to accept GM shipments, the global rice race hangs in the balance.

At the same time, rice, as a tropical crop, is increasingly suffering under climate changes that subject it to harsher and less predictable growing conditions and greater disease pressure. Genetic modification could help to improve pest and disease resistance as well as yield, say biotech promoters. Looking ahead, China is poised to commercialize GM rice, which has performed well in trials there, and traces of unapproved GM rice have already been identified across Europe in shipments from China as well as the United States. One way of thinking when it comes to GM rice is *as China goes, so goes the world*. But China, and even the major GM multinationals, are stepping lightly when it comes to commercializing GM rice. China has returned repeated requests for commercialization of GM rice back to the drawing board for further testing. A minor fiasco occurred in 2006 in which an unapproved variety of Bayer's Liberty Link herbicide-resistant rice was found across Europe, but it was quietly brushed under the rug with a retroactive approval—a moot point for future actions since the strain had not been cultivated in trials since 2001.

Thus, the unspoken undercurrent in the great GM papaya debate in Thailand was perhaps as much concern over the prospect of GM rice as an internal debate about tinkering with the nation's papaya crop. The real fear behind Thaksin's dramatic GM policy U-turn—which was quickly and quietly dismissed within a month—might have been more about Thailand's future as a rice exporter than about foreign viral genes turning up in *som tum* papaya salad. Indeed, it was rice exporters who had voiced strong opposition to the proposed legislation to allow GM crops to be commercialized in the country, and it may indeed have been the strength of their voices that prompted Thaksin's quick revocation of his proposal to parliament in the aftermath of the papaya episode.

While the GM papaya incident of 2004 stirred debate in the country and put Thailand on the GM map, the fast-track trajectory that debate over GM food in Thailand appeared to be shuttling down came to a screeching halt two years later. While a pro-GM stance remained an unspoken, if media savvy, consideration of a free trade agreement with the United States, and negotiators from both countries had continued to meet to iron out details of such an agreement, it was not to be—at least not on Thaksin's watch. Before such a trade treaty could be agreed upon, on the evening of September 19, 2006, troops descended on Bangkok and announced that the city and surrounding region had been placed under military rule. The constitution was suspended, and the cabinet, parliament, and constitutional courts were dismissed. Thaksin never got around to signing a free trade agreement with the United States before he was suddenly ousted, and the issue of GM approvals appeared to fade quietly in the face of more pressing national crises.

Just days before the coup, Greenpeace campaigner Patwajee Srisuwan had been acquitted on theft and trespass charges in the papaya field station trespassing case, and debate over GM papayas shrank into the background as the country was besieged by the aftermath of government instability. While the great GM papaya debacle had perhaps put Thailand on the GM map, focus on the future of biotech crops in Asia would move on to other shores.

CHAPTER 8

Got Hormones?

Engineering the Nation's Milk Supply

Long before the first GM food hit the shelves and far before "Frankenfood" frenzy erupted in Europe, transgenic technology had made its way into one of America's most sacred and wholesome foodstuffs: milk. Feeding off the 1980s biotech blockbusters, symbolized by the development of recombinant insulin in the human drug realm, several large multinationals—including Monsanto, Upjohn, Eli Lilly, and American Cyanamid—had begun to dream about making bacteria pump out hefty doses of a recombinant growth factor called bovine somatatropin (rBST—also known as recombinant bovine growth hormone, or rBGH). It was rumored that when fed to cows, this factor would dramatically increase milk production. Company executives reasoned that large vats of this high-octane hormone could spell millions in profits—$500 million, to be exact, based on preapproval anticipated sales figures reported in the *Wall Street Journal* in 1989.

Tampering with a "Sacred Cow"

Indeed, the genetic recipe for rBST had already been patented in 1980 by Genentech, the company that was about to put biotech venture capital on the map with Humulin, the recombinant human insulin drug that would debut in 1982 as the first recombinant human pharmaceutical to be approved by the Food and Drug Administration (FDA). But with its agricultural reverberations, rBST was an obvious piece of intellectual property that belonged in Monsanto's stable, and in 1981, the company had licensed the patent rights from Genentech. A few years after that, Monsanto began large-scale trials treating cows with this magic bullet that it hoped would greatly up the ante on milk production.

The potential to use naturally occurring growth hormones to increase milk production in dairy herds had been recognized as early as the 1930s, but the labor and expense of extracting the relatively small quantities of BST present in bovine pituitary glands made it prohibitive for wide-scale use. But five decades later, as the biotech bubble began to swell and new tricks and tools had come available, researchers revved their recombinant engines at the thought of churning out artificial hormones that might power the dairy industry. The potential to produce rBST on a massive scale and use it to supercharge the metabolism and milk production of America's dairy cows in a big way looked like an opportunity just begging to be milked.

By 1985, Monsanto had secured permission from the FDA to conduct large-scale field trials of the hormone, and cows reacted predictably, increasing their milk output on cue. But as with so many of their GM food products, Monsanto lacked foresight and failed to adequately, or perhaps accurately, consider consumer revolt as one potential outcome. Brashly building its rBST factory in Austria, Monsanto optimistically anticipated rapid approval and adoption in Europe. In the United States, the company touted the drug's supercharging ability to skyrocket production to increases of 25 or even 40 percent over already elevated levels of milk output in the country—forgetting that a growing glut of dairy surplus was driving prices down and hurting farmers at the time. Indeed, in those days, it seemed that almost everyone except for Monsanto was beginning to question whether the world, or at least the United States, needed more milk. Milk production in the country had steadily risen since 1975, while consumers had held their intake steady. That had led to an ever expanding government buy-back program, costing taxpayers billions in purchase and storage costs for the surplus.

In 1985, the same year in which Monsanto trooped out its field trials of rBST, the Food Security Act was adopted by Congress in an attempt to diminish the price supports that were drowning the program financially. Dairy herd sizes were cut, often dramatically, through such means as culling of overproductive herds. While Monsanto naively assumed that the U.S. obsession with growth at any cost would be a vision that dairy farmers shared, for many—particularly smaller independent dairy farmers—such a supposition shot fear rather than enthusiasm through their communities.

This was particularly the case in Vermont, a dairy-rich state that supplies almost half the milk for all of New England, and where until the last century cows outnumbered humans. With their smaller-sized farms and higher production costs, farmers there had been particularly hard hit by the leveling off of price supports in the face of rising costs. Vermont was, inevitably, also one of the states where Monsanto had contracted with university researchers to conduct safety and efficacy trials for rBST.

The practice of involving university scientists in commercial product trials, particularly in the pharmaceutical field, has increasingly come under scrutiny for obvious reasons. When scientists conduct research in the public domain, their investment in the outcome of an experiment is relatively limited. In such a context, science is practiced as a portal to increasing objective knowledge about the world, and researchers are inclined to publish their results, regardless of outcome. But when private interests fund and thus indirectly control the research process, bias can creep in. At best, experiments can be abandoned and questions left unanswered if the results of such forays threaten to turn up incriminating evidence.

At worst, critical components of the scientific method can be compromised, with data doctored or deleted and important results suppressed. In such enterprises, when companies control or conduct reviews of research prior to submission for publication, the "publish or perish" mandate so often present in academic science can be turned on its head. Documented cases of research bias in public-private academic collaborations in the human drug realm are increasingly being uncovered. Anecdotal allegations that scientists participating in safety and efficacy trials for Monsanto's rBST were encouraged to shift or suppress their research questions in a certain direction or lose favor and funding from the company have made their way into the grapevine of gripes against rBST.

Whistleblowers and Suppressed Evidence

At the other end of the spectrum are the rare scientist-activists who, when presented with data that disturbs them, decide to take their crusade to a personal level, potentially damaging their reputation and opening them up to charges of bias in the opposite direction. Sometimes, their commitments to their cause turn out to have been harbingers well heeded. Such was the case with Rachel Carson, the doyenne against DDT, or Stanley Prusiner, the proponent of prions as the causative agent of mad cow disease, who was once shunned but went on to win a Nobel Prize. But in other less celebrated cases that the world seldom hears about, such scientists are moved slowly to the sidelines, relegated as renegades to objectivity and the scientific method and often driven out by the detractors they targeted. It did not take long for this scenario to develop around the approval of Monsanto's rBST.

The first of these whistle-blowers, who would follow the latter path into scientific oblivion, was a Chilean scientist named Maria Lyng who worked in the agricultural school at the University of Vermont (UVM). In a news article that appeared in the British magazine *New Scientist* in early 1992, Lyng was reported to have been dismissed from her research post at the university after "asking awkward questions about the effects of BST." Lyng's research had focused on identifying the genetic basis for stillborn and aborted calf fetuses, and in her work, she had asked to obtain data and samples from the rBST studies, which she had heard were in some cases turning out aborted and deformed calves. Shortly thereafter, she was fired from UVM.

But Lyng did not leave quietly. Instead, she passed on critical data she had obtained relating to the rBST trials at the university that ultimately would provoke controversy. In 1991, Robert Starr, chairman of the agricultural committee of the Vermont state legislature, had requested data on the rBST studies performed at the university. They had been handed over by scientists, but not before the identifying numbers had been removed from the data, obscuring any conclusions that might be made between rBST and birth defects in a cow's progeny. When pressed, the university conceded that Monsanto had made them do it. The research contract between the university and the company clearly spelled out Monsanto's right to veto the release of any data for a year following the conclusion of the study.

But Lyng, disgruntled and convinced that there might be a relationship between rBST use and miscarried and deformed calf fetuses, was able to obtain the data with the identifiers. She passed the information on to a local activist group, Rural Vermont, as well as to the Vermont House and Senate agricultural committees.

Monsanto Concedes on Mastitis

After receiving the data, Rural Vermont commissioned a report by Andrew Christiansen, a state representative and an active member of the rBST debate through his membership in the state's House agricultural committee. The data were a bit fishy, but however they were analyzed, a significant percentage of the calves from rBST cows or their progeny appeared to be victims of birth deformities. Their abnormalities included a "bulldog"-type dwarf fetus that was aborted at six months; a "dipygus" calf possessing a double pelvis and extra legs, which caused difficulties during birth leading to the death of the mother; and an "encaphalocoele" fetus born to an untreated daughter of a rBST-treated cow that developed a large fluid-filled cavity in its head. These were gross and obvious abnormalities, not the kind often seen in short succession, if ever, in a dairy operation. Certainly the odd birth with a twisted leg or misshapen hoof might turn up every now and then on a farm, but the severity and phenotype of these abnormalities were, well, just that: far from normal.

The data appeared to associate more than severe birth defects with rBST treatment. There were also more minor yet still significant health ailments that plagued treated cows in the study. The report commissioned by Rural Vermont indicated the presence of various problems in treated cows including increased incidence of uterine infections, hoof rot, foot and leg injuries, and ketosis, a condition that is characterized by partial anorexia and depression and is accompanied by the presence of byproducts of fat breakdown, called ketones, in the animal's milk and urine. A detailed analysis of the data by David Kronfeld, an agricultural and veterinary expert at Virginia Tech who was commissioned by the Vermont state legislature, identified three statistically significant conditions in the rBST-treated cows: an increased incidence of retained placenta and ketosis, an elevated number of dead and deformed calves, and a higher number of "beefed" cows removed from the herd and sent to the slaughterhouse.

When the Rural Vermont conclusions were released, it triggered a news flurry, especially in key dairy states. Then two members of the U.S. House of Representatives asked the FDA to review the case and compare data submitted for review by Monsanto with the results uncovered by Rural Vermont. However, the numbers were small, and the way that Monsanto and UVM had reported the results of different studies was confusing. The data that had been reviewed by the FDA was different from the data from experiments leaked to Rural Vermont, even though all of the experiments had been conducted at UVM on behalf of Monsanto.

What was clear, though, however the data were interpreted, was that the cows given the experimental rBST treatment had a higher likelihood of health problems, including a fourfold increase in the frequency of mastitis, an infection of the udder, and a lesser likelihood of reproductive success. Following this revelation, after previously pleading no difference in bovine health relating to rBST treatment, scientists from UVM and Monsanto made a joint public confession in an article in the *Journal of Dairy Science* that in their experiments, rBST treatment did significantly increase the occurrence of mastitis in dairy herds.

This had important implications both for the health of the herd and because mastitis is usually treated by intramammary antibiotic infusion, which is a direct flushing of antibiotics into the udder. This in turn prevents that cow from continuing to produce commercial milk supplies for a set period after treatment; affected individuals must be marked in some way to prevent their milk from entering the holding tank. If antibiotic-tainted milk accidentally enters a supplier's milk shipment, the producer risks fines or losing its permit to ship Grade A milk.

From a health standpoint, antibiotics are just one implication of mastitis; cows with mastitis typically produce clotted or flaky milk and accumulate large amounts of pus at the site of infection. (While public aversion to antibiotics in milk is high, consumer sentiment surrounding pus in milk is even more unrelenting.) Later, once rBST had been approved and was in widespread use in the United States, the animal rights activist group People for the Ethical Treatment of Animals (PETA) would install a series of billboards mocking the dairy industry's "got milk?" campaign, inquiring instead if milk drinkers had "got pus?" in their rBST-sourced milk.

Of course, mastitis is not limited to cows being given rBST injections. Any cow can fall prey to mastitis, especially if herd hygiene is not adequately managed or if milk capacity dramatically increases. This was one of Monsanto's arguments against a direct link between rBST treatment and mastitis: Cows treated with rBST produced far more milk than their untreated counterparts, and thus one would expect to see a higher incidence of mastitis, due not necessarily to the drug but to the mere fact that they were producing more milk—a known risk factor for mastitis in the dairy industry. But the argument was circular, and experimentally, the incidence of mastitis in the group treated with rBST was statistically so much higher than the untreated control group that Monsanto had to concede the link.

Today, package labeling for Posilac—the trade name that rBST is marketed under—is required to contain the indisputable warning "Cows injected with Posilac are at an increased risk for mastitis (visibly abnormal milk) and may have higher milk somatic cell counts. Have comprehensive mastitis management practices in place on your dairy before using Posilac."

From Mastitis to Human Health Concerns

Animal health was one matter, but human health was another, more important consideration when it came to rBST. While the Vermont debate was raging, another unanticipated interloper had arrived on the scene in the form of Dr. Samuel Epstein, a professor of environmental and occupational medicine at the University of Illinois at Chicago. Epstein was a cancer expert who had been a key witness in the congressional testimonies leading to the banning of several harmful chemicals, including the pesticides DDT, Aldrin, and Chlordane, in the 1970s. That record alone should have been enough to make Monsanto quiver. But what called for an all-out assault by Monsanto was what Epstein now held in his hand: a black box of scientific supposition that drinking milk from cows treated with rBST might cause, in his opinion—gasp—cancer. What made his claim worse was the fact that this nebulous naysayer Epstein seemed to have appeared out of nowhere.

Samuel Epstein's Crusade

Epstein claimed to have gotten his start in March 1989 from the phone call of an anonymous insider, a farmer participating in the Monsanto trials who had grown concerned when his cows became sick. "If it makes my cows sick, their milk will also make people sick," the farmer was reported to have told Epstein, urging him to dig more deeply into the matter. And dig he did. What Epstein unearthed about the potential risks of rBST made him profoundly uneasy, enough so to make him adopt opposition to rBST as his personal crusade.

Epstein began his assault with a letter to the FDA in July 1989, which he also forwarded to a number of members of Congress, documenting what he felt were significant health concerns that could stem from approval of rBST. In his extensive letter, Epstein provided a blow-by-blow critique of the studies that had been performed en route to potential rBST approval. At the time that Epstein began his investigation, evidence of increased frequencies of mastitis uncovered by the UVM studies had not yet been made public; Monsanto was still insisting that there were no health effects for cows treated with rBST. Epstein had analyzed data from rBST treatment experiments conducted in Pennsylvania by Monsanto's competitor Cyanamid, and in Missouri by both Cyanamid and Monsanto, which indicated increased incidence of mastitis in treated cows.

One risk that Epstein pointed to was the potential human health effects of increased antibiotic exposure resulting from drinking milk from mastitis-infected cows. But the centerpiece of his argument was that treatment of cows with rBST led to higher levels of another growth factor, known as insulin-like growth factor 1 (IGF-1), in treated cows' milk. As Epstein pointed out, elevated levels of IGF-1 in humans had, in some studies, been implicated in growth disorders in children as well as breast cancer in adults. Although the amino acid sequence of BST from cows and humans was sufficiently different such that bovine BST was not active in humans, IGF-1 was a different story. Versions of this hormone from cows were known to cross-react in humans, potentially mimicking the tumor-promoting and growth-regulating functions of the endogenous version of this factor in humans. Thus, drinking milk from rBST-treated cows held the potential for exposure to higher than normal levels of bovine IGF-1, and perhaps dangerous health effects, according to Epstein.

Epstein was not satisfied with the agency's response to his letter, so he decided to take his crusade public. A week after his letter to the FDA, the *Los Angeles Times* published an Op-Ed piece authored by Epstein titled "Growth Hormones Would Endanger Milk," which distilled the key points he had made in his letter to the FDA. Not surprisingly, his article was followed up by a defense from the FDA, which was printed in the paper five days later. Epstein's piece had been intended "more to frighten than enlighten," wrote Gerald Guest, the director of the FDA's center for veterinary medicine, and had contained "some factual information and a great deal of scientific exaggeration." Furthermore, Guest reassured readers, "The men and women at FDA are not dumb. The food safety scientists at FDA are amongst the most knowledgeable and capable in the world today, and we do care about the people we serve, the consumers of meat, milk, and eggs."

The soothing, slightly pedantic tone of Guest's message perhaps came too late. Epstein's genie had escaped from its bottle, and the hypothesis that elevated levels of growth factors in milk from cows treated with rBST might cause cancer and other detrimental health effects had made its public debut and would live on for decades. Despite the distraction that Epstein had invoked—including the subsequent publication of a scientific article entitled "Potential Public Health Hazards of Biosynthetic Milk Hormones" the following year in the *International Journal of Health Services*—the FDA nonetheless stayed its course, supported by Monsanto's continued enthusing over the drug and a barrage of defenses against its detractors.

The FDA Under Fire

What happened next, and how, is complicated and the subject of some controversy. First, a veterinarian named Richard Burroughs came forth, contending that he had been fired from the FDA for "slowing down the approval process" of rBST. "It used to be that we had a review process at the Food and Drug Administration. Now we have an approval process. I don't think the FDA is doing good, honest reviews. They've become an extension of the drug industry," Burroughs told the *New York Times* in early 1990.

The rBST accusation added to a growing list of complaints against the agency. Vermont Senator Patrick Leahy, the chair of the Senate Agriculture Committee, commanded the congressional General Accounting Office (GAO, later called the Government Accountability Office) to commence an investigation. This was, perhaps, a natural extension of another GAO investigation that had begun two years earlier and had found, as stated in the title of the initial GAO report, "FDA surveys not adequate to demonstrate safety of milk supply."

In the sixty-eight–page GAO report on rBST triggered by Burroughs's accusations, the authors concluded that the FDA had addressed the major critical FDA review guidelines for rBST in three areas: human food safety, animal safety, and drug efficacy. Despite such compliance, they found a fourth critical consideration that the FDA process had overlooked: an investigation into "the indirect human food safety risks that result from animal health effects caused by the use of the animal drug." The report expressed concerns about elevated antibiotic levels in milk, which it suggested might "already be too high from present antibiotic usage and how well these levels are monitored," referencing a second report the GAO had issued earlier that month reemphasizing the need to develop a more effective FDA strategy for addressing animal drug residues in milk. On the basis of this important oversight, the GAO recommended, in the title of its report, that "FDA Approval Should Be Withheld Until the Mastitis Issue Is Resolved."

In response to this criticism, the FDA assembled its Veterinary Medicine Advisory Committee to hold hearings that were open to the public. Many anti-rBST activists came to testify, as well as a slew of scientists, FDA officials, company representatives, and members of animal-rights organizations. In the end, though, the committee decided that the FDA had newly instituted adequate steps that would make the risk of antibiotic residues in milk from rBST-treated cows manageable, and that therefore, approval of the drug should proceed. As far as the FDA and Monsanto were concerned, the mastitis issue had been resolved.

Although those who testified before the Veterinary Medicine Advisory Committee had presented a range of views and had come armed with their own, often contradictory, data, the blanket decision that rBST should be approved angered those who had sided against it. Taking a new tack, they accused the government agency of failing to properly monitor and respond to Monsanto's illegal engagement in preapproval marketing of the drug as "safe and effective" prior to FDA review and approval—an activity specifically prohibited by federal law.

Monsanto had, the critics argued, reeled in farmers by paying them to attend pro-rBST focus groups, issuing a promotional video in collaboration with the American Medical Association, and making a public presentation promoting the drug at Louisiana State University, all prior to final FDA review and approval. Vermont's Bernie Sanders (at the time a member of the U.S. House of Representatives and now a senator) called on the Office of the Inspector General to investigate these claims. Again, the report found no major faults, although the reviewers concluded that the university presentation appeared to fall into a hazy category of conduct. Opponents of rBST were running out of options. rBST appeared to have arrived to stay.

To Label or Not to Label, the rBST Question

On November 5, 1993, the FDA granted the landmark approval of Posilac, Monsanto's proprietary formulation of rBST, marking the first time that a food derived from the use of a genetically engineered drug had been allowed to enter the food system. As David Kessler, then commissioner of the FDA, expressly proclaimed, "There is virtually no difference in milk from treated and untreated cows." Accordingly, the agency concluded that when such milk arrived on grocery store shelves the following year, no special labels would be required to mark the product that had been the subject of so many years of research controversy and government scrutiny.

Indeed, anticipating such a debate, the FDA recognized the need to explicitly address the subject of labeling. Just five days after Posilac went on sale to the public, the FDA released its "Interim guidance on the voluntary labeling of milk and milk products from cows that have not been treated with recombinant bovine somatotropin." Because no milk could be deemed to be "BST-free" due to the endogenous presence of a cow's own BST in its milk, labels advertising "BST-free" would definitely invoke false advertising, according to the FDA. They suggested, in its place, labels along the lines of "from cows not treated with rBST." But even that statement could be misleading and confuse consumers, the agency asserted. Consumers might put on their thinking caps—backwards, in the mind of the FDA and Monsanto—and assume that such milk might in turn be safer or of higher quality than milk from untreated cows.

Thus, labels providing information regarding the rBST status of cows from which milk was derived also needed to have a statement putting that information in "proper context." Such proper context, in the words of the FDA, would consist of a statement such as "no significant difference has been shown between milk derived from rBST-treated and non-rBST-treated cows." The FDA further suggested that each state should individually examine labeling and advertising claims related to rBST status, because states had historically taken the lead in overseeing milk production. Where available, data pertaining to consumer perception could be used to determine whether a particular label might be deemed misleading or not.

Putting the Labels to the Test

These guidelines were only "interim guidance" and were, of course, nonbinding. But it did not take long for them to be put to the test. Not surprisingly, given its history of tussles in the rBST trials, Vermont was the first state to step forward. In April 1994, Governor Howard Dean signed into law a mandatory labeling scheme, in which any product created from milk from rBST-treated cows would be required to bear a label stating that fact. The law had roughly a year before it was to go into effect, and industry was quick to respond. A lawsuit was filed on behalf of the International Dairy Foods Association, asserting that the law violated corporations' first amendment rights not to speech but to silence. At first, it looked like the legality of the labels—in some cases mere blue dot stickers applied to products or price tags on store shelves—would stand, as a U.S. District Court judge turned down the industry groups' request to block implementation of the law. But then, on appeal, the Court of Appeals for the Second Circuit reversed the initial finding, deciding that the consumer's right to know neither superseded the right to silence nor justified the need to label.

Next up on the labeling law docket was Ben & Jerry's, a famous Vermont company. Ben & Jerry's, a staunch anti-rBST crusader, took on an anti-labeling law in Illinois, a state that had opted to pass guidelines forbidding any labeling relating to rBST status of milk products. As a result of this Illinois law, regulators required that cartons of Ben & Jerry's products containing statements pertaining to rBST be removed from store shelves in the state. The case was settled out of court, and Ben & Jerry's prevailed, pushing forward with products displaying labels and paving the way for producers in Illinois to follow suit.

The labeling debate calmed down for a while, but that did not mean that labels were not in place on products. The FDA guidelines and resulting lawsuits seemed to have settled the matter: Labels pertaining to rBST status would be allowed, but only if they contained express disclaimers of any superiority or benefit. Specific legislation could be carried out at the level of individual states. Consumer demand as well as organic guidelines kept dairies labeling, at least in some states and cases.

It was not just milk drinkers who had their eye on rBST labels. Monsanto kept a close watch on which dairies were saying what about rBST on their cartons and packages, lying in wait for cases that might contradict what the FDA guidelines and legal decisions had established as acceptable. Most labels were within bounds, but then in 2003, a warning buzzer sounded at Monsanto headquarters in St. Louis. Label watchers had spied a case ripe for investigation.

Monsanto vs. Oakhurst

The label in question was borne on every one of the milk-related products sold by Oakhurst, a small dairy in Maine that had been continuously operated by three generations of the Bennett family since 1921. "Our Farmer's Pledge: no artificial growth hormone used" read the offending statement, plain and simple. But Monsanto begged to disagree, arguing that the message was anything but simple. "We believe Oakhurst labels deceive consumers; they're marketing a perception that one milk product is safer or of higher quality than other milk," Jennifer Garrett, a spokesperson for Monsanto, told the *Portland Press Herald* in an interview.

To back up its assertion, Monsanto pointed to consumer research that the company had commissioned from MSR Research Group (Market Street Research), a company based in Northampton, Massachusetts. The research found that a majority of Oakhurst consumers considered the labeled milk healthier or safer to drink than that without labels. A court date was set for early the following year.

Before the case came before the court, the parties settled, with Oakhurst agreeing to place an addendum on its label conceding that "FDA states: No significant difference in milk from cows treated with artificial growth hormone." Labels were still allowed, but with the Oakhurst case, Monsanto let the nation's dairy farmers know that big brother was watching what they said about Posilac, their prima donna poster child for the impending GM revolution.

rBST Meets Defeat in Europe and Canada

While the company continued to exert pressure, Monsanto was wise not to push for an outright ban on labeling. After all, the FDA had issued guidelines allowing some sort of acknowledgment, despite the burdensome need for disclaimers. And a vast majority of U.S. consumers wanted any genetically modified food product to be labeled as such.

A 2001 survey by the Pew Charitable Trusts's Initiative on Food and Biotechnology—a reputable research organization in the agricultural biotech arena—had revealed that despite the fact that nearly two-thirds of Americans believed they had not eaten genetically modified food, three-quarters of the population surveyed said "it was important to them to know whether a product contains genetically modified ingredients." A whopping 46 percent responded that it was "very important." Consumer demand kept the dairies labeling, and by 2002, eight years after its approval, use of rBST had caught on in only 22 percent of the nation's dairy cows, according to statistics provided by the U.S. Department of Agriculture (USDA). Outside the United States, rBST had been soundly rejected, based on concerns for the health and welfare of animals treated with the recombinant hormone.

In October 1999, the European Union had issued a press release announcing its decision to ban the use of rBST within its bounds from the beginning of the following calendar year. Canada had issued a similar ban that same year, but not before purported offers of bribes from Monsanto to regulatory scientists in that country had been revealed. Testifying before the Canadian Senate Standing Committee on Agriculture and Forestry, Canadian government scientists had told of stolen files from office cabinets, offers of several million dollars in payoffs from Monsanto for signing off on rBST approvals, and pressure and coercion to sign off on drugs of "questionable safety, including rBST." In the end, Canada's Health Protection Branch had said no to rBST in the wake of this public relations fiasco and had denied regulatory approval to Nutrilac, the name under which Monsanto hoped to market rBST in Canada.

While the more humane European and Canadian regulatory agencies may have balked at the health implications for cows treated with rBST, the conditions arising from rBST treatment seemed minor in comparison to some of those inflicted by the factory-farming approach that was standard in the United States. It would take far more than a few udder infections to make Americans blink at the use of rBST. But blink they would, though it would take time and effort for activist groups to educate them about what they considered the potential health concerns laying in wait.

The Making of an Activist and the Birth of a Grass-Roots Movement

It was not long after the Oakhurst labeling fiasco in Maine that Rick North first learned about recombinant hormone use in the dairy industry. He would go on to become an unexpected activist and perhaps the ongoing drama's most effective anti-rBST campaigner. When one hears the term "anti-biotech activist," the image of North is not what typically comes to mind. With his pressed khaki pants, navy blue knit polo shirt, middle-aged demeanor, and happy, big-toothed grin, he's the kind of guy you'd more likely imagine crossing paths with on the golf tee or seeing on Sundays at church. In fact, it was through church that North first learned about rBST.

In 2002, North was doing research for a church program on how eating habits can affect health. In his words, he "stumbled across this information on GM food that looked kind of concerning, that I'd never heard of before, specifically on rBST." North learned that a possible connection between rBST use in dairy cows and an increased rate of cancer in humans resulting from drinking milk had been hypothesized, and that was when he decided he had to delve in further. After all, North had recently retired from a twenty-one-year career with the American Cancer Society, spending his last five years as director of the organization's Oregon chapter. Moreover, each of his wife's three sisters had suffered from cancer, and he was concerned about her potential genetic risk as well as that of their children.

As he tells it, "We made a very easy decision, in about two minutes after my wife was done reading, and that was the last day that we bought rBST milk." He figured that would be the end of it. "This was not a major part of the church program at all, and I had no idea that I would go on to lead a campaign educating the public about rBST," North says in retrospect. His personal interest in the issue continued, and when a state ballot measure proposing mandatory labeling of genetically modified foods was put forth in Oregon in the fall of 2002, North was quickly recruited to join the executive committee. That initiative, Measure 27, ultimately failed when it came to a vote, despite initial voter enthusiasm and after biotechnology trade groups and corporations, including Monsanto, had pumped more than $5 million into fighting its passage, making history as one of the state's most expensive ballot measures on record.

It was through his involvement with Measure 27 that North met a physician by the name of Martin Donohoe, a local doctor and lecturer who was on the advisory board of the Oregon chapter of Physicians for Social Responsibility (PSR), a nonprofit organization active in opposing nuclear weapons and promoting environmental health issues. In the wake of the disappointing, and—in the minds of some, rigged—failure of Measure 27, North and Donohoe agreed that this was too important an issue to let go. "If I can raise the money, we can start a program," North told Donohoe. After several appearances before the Oregon PSR board, North got the green light. "We raised the money and away we went," he says, making it sound easy in retrospect. "We just thought the public needed to know about this."

But there were no models or organizations in the country that had taken a lead on this issue, and public opposition had more or less died down by that time, so North had to chart his own course. He started with the science, gathering concerned doctors, scientists, and other activists into his fold. They decided not to go to the FDA, figuring it was a lost cause. And the idea of a lawsuit was out because "that just wasn't us," North recalls.

Instead, North and his small band of organizers began a three-pronged approach, beginning with making public presentations to groups and businesses—the Rotary club, moms' groups, high school and college classes— any public group that was willing and open to hear them out. The second arm of the strategy was to call up dairies that were using rBST. At first, they called just the local dairies in Oregon and told them, "Hey, we've got a problem with this, and could I come and show you what our problems with it are?" Some of the dairies cautiously invited North in, others refused, and some didn't return his phone calls. North didn't let that deter him. "I mean, put yourself in their shoes—here's some guy coming and telling you that the products you're selling aren't safe for human health, or at least there's a risk. I'm not going to be number one on their hit parade," he recalls. "There was a lot of rejection involved, but I thought, so what, I'm not going to convert everyone on the spot, and it was all very civil."

While he was busy giving talks and telephoning dairies, North was also hard at work on the third prong of the strategy. This was an experimental approach that in the end turned out to be a much more powerful weapon than he might have imagined. North printed up a series of "these crazy little postcards, that are very polite, very low key and respectful, with printing on the back saying we've got problems with this, we're concerned that you use rBGH, and would you please consider stopping its use and labeling so the public would know?" On the front, North asked attendees at talks and event tables to sign their names and print their addresses, if they were willing to send a message to a particular dairy.

With the postcards, "We didn't single out any one dairy, because we wanted them all to stop; we pretty much just chose them all," North recalls. People started sending in the signed postcards, at first hundreds and hundreds, later thousands and thousands. "We had no idea—there was no template or guidelines for doing a campaign like this, and we just thought, okay, we'll try the postcards. We had our basic information sheets, later we printed up pretty brochures to hand out, and we had information on our website, but this was really just low-tech stuff."

A Breakthrough at Tillamook

The first big breakthrough was Tillamook, a cooperative dairy out on Oregon's coastline and the nation's second largest supplier of chunk cheese. According to North, Tillamook had already done some market research of its own, and the company was finding that "wow, people didn't like this stuff! And then they were getting hundreds and hundreds of postcards." Soon, the Tillamook board voted unanimously to go rBST-free with its cheese, a milestone for the campaign, despite the fact that North hadn't made personal contact with the dairy. "At the time, I couldn't get to first base with Tillamook. They wouldn't return my phone calls," North admits. But they had said no to rBST nonetheless, and the postcards had likely played an important role.

No sooner had the board's decision been announced than alarm bells rang in St. Louis again, and within days, a Monsanto representative had arrived on the scene. Tillamook's cooperative organization structure allows an override and general membership vote on an issue if enough members sign a petition stating they oppose the board's decision. Monsanto saw this as a golden opportunity to intervene. According to North, Monsanto representatives drafted a petition, arranged for the requisite number of farmers to sign it, and a Monsanto representative actually delivered it in person to Tillamook headquarters.

Tillamook's president was furious, and a local TV reporter had taken on the story—sure to strike controversy—as a personal, breaking news crusade. North still couldn't get in the door at the dairy, but he was satisfied to watch from the sidelines as co-op members voted 83–43 to back the board's decision to go rBST-free. As North wrote afterward in an E-mail update to his supporters, "It's funny, but regarding what Oregon PSR will do now, it really didn't make any difference which way Tillamook voted. We'll just keep developing our grass-roots efforts and continue to educate the public so they can make well-informed decisions about buying dairy products. In many ways, we're just getting started." Indeed, he was right. His work had just begun.

Tillamook's rBST ban began on April 1, 2005, but it was no April fool as the decision sent a shock wave through the dairy industry. Besides being the country's second largest producer of chunk cheese, Tillamook was well-respected, and setting such a precedent would be a hard act not to follow. Tillamook had even publicly denounced Monsanto—an action few had been brave enough to attempt—and the dairy appeared to have survived, so there were few excuses to stop others from following suit.

And follow suit they did. Tillamook's decision triggered a mass action. "We were having one dairy after another go rBST-free in Oregon and Washington," North enthuses. "All the postcards, public pressure—we had people e-mailing and phone calling too, it was a grass-roots effort." In the aftermath, it was revealed that Tillamook had received more than 6,500 public comments during the fray, with 98 percent of those expressing opposition to rBST.

The rBST-Free Movement Spreads

Following that victory, North stepped up his campaign to include outreach to institutions such as schools and hospitals that contracted for large volumes of dairy products, which was a natural move for PSR. By switching contracts to dairies pledging to go rBST-free, institutions could exert massive pressure. A parochial school in Oregon threatened to make a move, and a local dairy went rBST-free. Going rBST-free was beginning to be good for business. The near-defunct Schultz Creamery, located in Bismarck, North

Dakota, was revived under the management of Great Plains Dairy Partners after cutting an exclusively rBST-free deal with Bon Appetit Management, a major institutional food service provider. Schultz was able to hire back ten employees with the resurgence of its Deja Moo brand of rBST-free milk.

Then the movement began to spread to dairies across the country. By the end of 2006, dairies in New York, Texas, New Jersey, and Montana had joined those in New England and the Northwest in staging all-out rBST bans, and it was rumored that Wal-Mart, Kroger, and Dean Foods (the nation's largest dairy company) were all investigating processing or buying rBST-free milk supplies to meet consumer demand. Healthcare giant Kaiser Permanente and Catholic Healthcare West hospital systems were on the move to implement rBST-free sourcing as well.

Starbucks Goes rBST-Free

If Tillamook had been North's first big breakthrough, his second was soon to come. At the end of 2006, Starbucks—the world's number one retailer, roaster and brander of coffee—began considering whether to take all its shops rBST-free. Public sentiment was strongly leaning in that direction, but North was fortuitously afforded an opportunity to further brew that idea along with the company. Coincidentally seated next to a Starbucks executive at a sustainable food association dinner in late 2006, North "gave him an earful." It was in many ways the chance he had dreamed of. "There's a whole lot of lattes being sold out there! Starbucks has a tremendous milk usage," he points out. Before he knew it, North had finagled a meeting at Starbucks headquarters in Seattle, where he spoke to a contingent of fifteen or twenty staff members. After three hours of talking, he left, determined to continue contact with the company and keep sending them information about the campaign. You don't convert everyone on the spot, as he had said when he first started his campaign.

Lo and behold, a few weeks later, his phone rang, and when North picked up the receiver, it was Starbucks on the other end of the line. Could he come back and make his presentation again, only this time to their dairy suppliers from all over the country? North was in a state of disbelief. "I thought, OH MY GOD! This is an opening like I will probably never see again! This was definitely the big time!" So he headed up to Seattle once again, showed his PowerPoint presentation to the dairy suppliers, and spent another three-hour session fielding questions and answers. The atmosphere was cordial, but when he left, not everyone was smiling. "Again, here I was, telling all of these dairy suppliers, none of whom were rBST-free, things that obviously were making some of them quite uncomfortable. They were getting the health side of this that they don't get from Monsanto or the FDA."

The discussion with the dairy suppliers continued after North left, and in the weeks following the meeting, North continued to send information Starbucks's way: a news story here, clippings from around the country there, updated health information on the potential links between cancer and drinking rBST milk. Starbucks was talking with Monsanto, North knew, and his grass-roots leverage paled in comparison to the corporate big guns Monsanto had to draw upon. But sometimes, the grass is greener on the other side. To his amazement, North prevailed. Starbucks made its decision: The company would go rBST-free in all of its U.S. stores by the end of 2007.

Slowly, region by region, Starbucks pulled rBST milk out of every latte, every mocha, every frappuccino the stores served. "We pride ourselves on the understanding we have developed of our customers' desires based on our close relationship with them. We took this step exclusively in response to continued customer demand for dairy produced without the use of rBGH," read a press release issued by the company in early 2008, after the conversion was complete.

Chipotle Leads the Way

In the meantime, the national chain restaurant Chipotle, majority-owned by McDonald's, had announced on November 5, 2007 that it would beat Starbucks to the punch. "First and Free: Chipotle Mexican Grill Is Nation's First Chain to Go Entirely rBGH-Free" read the press release. In it, Chipotle quoted results of a recent survey conducted by the Opinion Research Corporation that found that 81 percent of respondents preferred to purchase dairy products from cows not treated with synthetic hormones, assuming the price difference was small or nonexistent. Of those who had no preference with respect to hormone treatment, half said that they would change their mind and avoid such products if they knew that the hormone had been banned in a number of other countries.

North's public education campaign was apparently working, as 58 percent of survey respondents were by now aware that many dairy cows in the country were receiving rBST treatment. "As consumers become more aware of the issues associated with rBGH use and the alternatives companies like Chipotle are providing, they are clearly stating their preference for rBGH-free dairy products," North was quoted in the conclusion of the Chipotle press release. Earlier that year, he had been invited to give his PowerPoint presentation at the company's national marketing meeting in Chicago. It seemed that the rBST tide had finally turned.

But North knew better than to let his guard down. As he had said after the initial Tillamook victory in 2005, "in many ways, we're just getting started." Still, Bloomberg News had reported in October 2007 that profits from Monsanto's sales of Posilac were down in the dumps, projected along with selective herbicide sales to fall 16 percent in the upcoming year because of "pressure in the dairy business." In August 2007, Monsanto had filed an arbitration claim seeking damages from Novartis's Sandoz unit, the contracted manufacturer of Posilac in Austria. Between 2003 and 2006, failures in quality control tests performed by the FDA at the Sandoz manufacturing facility had caused a sudden downturn in the supply of Posilac, a reduction in availability to farmers, and a price increase for the product. In response, Monsanto had been forced to construct a new plant in Augusta, Georgia—an expensive undertaking that had only begun operations in 2006. Monsanto couldn't afford to lose ground with Posilac now.

The Labeling Battle Is Revived

And so, against these odds, Monsanto leaned on an old card it had tried time and time again and returned to the issue of labeling. No matter Starbucks's nationwide pledge. No matter the Chipotle billboards boasting hormone-free goods or the company's "Clean Livin'" radio ads with rocking ranchers singing "shooting up ain't cool" for their cows. Now that more consumers than ever were aware of rBST use in the dairy industry, and most wanted nothing to do with it, Monsanto entered into the ring for one last "don't ask, don't tell" attempt to ban rBST labeling.

Pennsylvania Moves Against Labeling

Company officials deny that they started the fight, but once the first punch was thrown, they were at the ready to step in. On October 24, 2007—as dairy after dairy across the country were signing on to go rBST-free—the office of Dennis Wolff, Pennsylvania's secretary of agriculture and the owner of Pen-Col farms, a 600-acre dairy cattle business with joint operations in Wisconsin and the United Kingdom, suddenly issued a controversial press release. The release stated that "false and misleading labels" relating to the absence of synthetic hormones in milk would have to go. As part of this campaign, sixteen dairies in Pennsylvania, New York, New Jersey, Connecticut, and Massachusetts had received warning letters that their rBST-free labels would be deemed illegal beginning January 1, 2008.

In the press release, Wolff suggested that consumer demand had driven his desire to do away with the labels. "We're seeing more and more marketing that is making it hard for consumers to make informed decisions," Wolff complained. "It's a subject the department continues to receive many calls about." Confusion aside, the rBST labels couldn't be trusted anyway, according to the department. With regard to rBST labeling, "There is no scientific test that can determine the truth of this claim," emphasized the press release.

A little digging revealed that scientific tests weren't required for labeling of other foodstuffs in the state. Michael Hansen, a senior scientist from Consumers Union and a longtime foe of rBST, argued that claims advertised on many food labels could not be proved in a laboratory. "Probably the most prominent example is reconstituted orange juice," he pointed out in an interview in *Supermarket News* magazine that December, as the deadline for the disappearance of the labels loomed. "Orange juice that's fresh-squeezed has to be labeled differently than if it's from concentrate. If it's reconstituted, then it has to be labeled. And I don't know of any quick tests to do on orange juice to tell if it's freshly squeezed or made from concentrate."

Moreover, it appeared that Wolff's conjecture that his ruling was in response to consumer confusion and demand also rang hollow. "Mr. Wolff's office could not provide surveys or research showing that consumers were confused by the issue, and was unable to come up with even one name of a consumer who had complained," wrote Andrew Martin in a *New York Times* article on the proposed change in labeling laws. Martin also pointed out the inconsistency of this about-face in the context of allowable labels in the state: "There are exceptions to his rule, for what he describes as 'puff' claims like 'farm fresh' and 'locally produced.' Isn't he saying that milk produced in New Jersey is inferior? And how do you scientifically prove it's from Pennsylvania anyway?"

Monsanto's Involvement—and the Public Backlash

And so one was left to wonder at how and why Pennsylvania's department of agriculture had suddenly made such a decision and if Monsanto wasn't somehow secretly involved. Monsanto had tried the overt approach with Oakhurst dairy. Monsanto had also pressured Steven Rowe, Maine's attorney general, to revoke that state's "Quality Trademark" seal program, which required that cows not be treated with rBST, and the company had received a slap in the face. After failing on both of those accounts, Monsanto had kept the pressure up in small ways—for example, complaining to the FDA and the Federal Trade Commission (FTC) earlier in 2007 to have rBST labels declared misleading. They had failed there too, with both the FDA and the FTC concurring that current guidelines allowed such labeling and that food companies had the right to inform consumers that they do not use rBST.

Thus, when the Pennsylvania case arose so suddenly and against the stream of consumer demand for rBST products, critics began to speculate. Kevin Golden, an attorney for the Center for Food Safety, wondered aloud in a *Supermarket News* article if "the company, which markets the hormones under the brand name Posilac, wasn't somehow involved with pressuring Pennsylvania officials into their decision." In March 2008, Andrew Martin of the New York Times was able to connect the dots. In an article published in the New York Times, he revealed that an advocacy group known as American Farmers for the Advancement and Conservation of Technology (AFACT) had been formed to advance "a counteroffensive to stop the proliferation of milk that comes from cows that aren't treated with synthetic bovine growth hormone." AFACT was supported financially by Monsanto, and the company had had a hand in helping to organize the group. Lori Hoag, a spokeswoman for Monsanto, was quoted in the article as saying of AFACT, "They make all the governing decisions for their organization, Monsanto has nothing to do with that." Regardless of Monsanto's direct involvement, the Pennsylvania rule was on the table, promising to fulfill Monsanto's wishes in removing a force that had proven detrimental to Posilac's acceptance and sales.

When Rick North learned of the Pennsylvania decision, he was prepared. The next day, he sent out an e-mail informing his constituents of this transgression. "This is about as serious as it gets," read his message. "The rationale given is that current labeling is 'making it hard for consumers to make informed decisions.' This statement can best be characterized as the brown substance that comes out of the back end of a male bovine . . . Obviously, Monsanto's cronies have been working hard to pressure Pennsylvania to take steps to stop the tidal wave of dairy processors going rBGH-free." In response, North was ready to unleash a tidal wave of his own. "Oregon PSR will do everything possible to counter this outrage," he ended his announcement.

Within days, the office of Pennsylvania Governor Edward Rendell was inundated with calls and e-mails expressing outrage over the decision. Whether they were confused or not, one thing was clear: Consumers just did not want artificial hormones coming anywhere close to their milk. "Any time you dicker with hormones and that, I don't trust it. Leave it the way Mother Nature intended," said a sixty-four-year-old shopper interviewed in the aisles of a local market by the *Pittsburgh Tribune-Review*. Consumer groups, the Pennsylvania Association of Milk Dealers, and the director of the Allegheny County Health Department all spoke out against the ban.

At first, Rendell pushed back the implementation date by a month, to the start of February 2008, to allow time for a full investigation. But the calls kept coming. In the end, more than sixty organizations signed a letter of opposition and thousands of citizens registered messages of protest. Rumors of lawsuits were circling. It was clear that consumer confusion could no longer be called upon as an excuse.

The chorus became too loud to ignore. While attention was focused on Pennsylvania, Kraft—the nation's largest provider of chunk cheese—announced that by the middle of 2008, it would transform its entire low-fat line of cheeses, consisting of thirty to forty different products, so that they would be rBST-free. Three days later, on January 17—two weeks before the Pennsylvania ban was to take place—Rendell's office issued a retraction. "The public has a right to complete information about how the milk they buy is processed," the governor's statement began. Labels would be allowed but restricted to the wording spelled out in the original 1994 FDA advisory. Opponents of the ban cheered. "PENNSYLVANIA—YES!" read the subject line of North's E-mail announcement early that morning. "This is a big victory for us," Consumers Union scientist Hansen said in the *Pittsburgh Post-Gazette* the following day. "This is a 178-degree turn. It's not quite 180 degrees, but virtually all."

The remaining two degrees of the turn came later that summer. On August 6, 2008, Monsanto issued a press release announcing its intention to divest its rBST operations and product. While such a move did not definitively spell the end to the hormone (since it was still approved for use), it did seem to be a last gasp in the face of outright consumer rejection. As *Farm Journal's* Ag Web blog wrote in a post entitled "Bye-Bye BST?": "It's clear that we have turned a significant page with this decision, and a new chapter in dairy will be written for here out."

Whether Monsanto's verdict marked the final failure in attempts to stem consumer choice with regard to rBST remained to be seen. But it was the bellwether of a move away from the passivity that had so often been assumed of the American consumer with regard to genetically modified food. It also sent a warning signal that in this land of plenty, what people ate was becoming a matter of personal passion as well as politics. Monsanto had been given a serious hand-slapping, and the first GM food fray on American soil had taken root.

CHAPTER 9

From Pharm to Fuel

The Future of GM Food

Monsanto had fought long and hard for more than two decades to protect its rBST product Posilac from consumer defeat. After all, Posilac had been the first GM food-related product to gain commercial approval in the United States, and as such, it held symbolic significance as the forbearer of a future GM revolution. But by the dawn of the twenty-first century, it was no longer cutting-edge technology to pump out pharmaceutical proteins in bacteria, as Posilac was produced. The biotech industry was hard at work, dreaming up so-called second generation GM products, which they hoped would wow the public more than Posilac had.

Your Food on Drugs?

One of the early entrants into this field of dreamers was a company called ProdiGene, a small start-up that had been spun out of Pioneer Hi-Bred International and recruited to set down roots in Texas during George W. Bush's governorship there. A cornerstone of the biotechnology revolution of the 1980s had been the ability to produce drugs, such as Posilac or Humulin (a recombinant form of human insulin) in bacteria. These tools had significantly lowered costs and upped production and could be protected by patents to boot. As such, they had drawn a huge crowd of followers in the corporate biotechnology realm. But now, new tools were needed, or at least new raw materials upon which to temper those tools.

Although small in size, ProdiGene was thinking big, even bigger than bacteria. Recombinant *E. coli* had afforded incredible cost savings and production scale. ProdiGene proffered that producing the same kinds of proteins in plants, grown in the good of nature and powered by free photosynthetic power, could lead to even greater gains. Food crops like corn were inherently constructed to churn out their proteins using only the power of the sun, so it would just be a matter of putting the proper DNA sequence in the plant and harvesting the proceeds from the fields.

ProdiGene wasn't just dreaming, though. Back in 1996, when the company was still a small start-up known as Terramed Inc., it had been awarded U.S. Patent No. 5,484,719, for "Vaccines Produced and Administered through Edible Plants." This was the first patent ever awarded covering proprietary technology that would allow plants to produce proteins that could be used as vaccines. This patent was essential to the company's arsenal as it sought to put plant-made pharmaceuticals (PMPs)—or biopharming, as it would later colloquially become known—on the map.

ProdiGene worked fast and starting in 1997 began to apply for regulatory approval to conduct field trials with corn engineered to produce a variety of secret, experimental drugs the company had devised. Between 1997 and 2001, the company was awarded ninety permits in nine states plus Puerto Rico to conduct field trials of corn (with one permit for tomatoes) expressing pharmaceutical proteins that were identified only as "CBI," for "confidential business information." As part of the regulatory guidelines, inspectors from the U.S. Department of Agriculture (USDA) were to visit the site of the field trials at certain times, and the company was to ensure adherence to rules regarding disposal and containment of the test crops.

It appeared that ProdiGene was poised to be one of the pioneers in the PMP game, but what first put ProdiGene in the media wasn't exactly what its investors had in mind. In the fall of 2002, a USDA inspector at one of the company's test plots, in Nebraska, came upon something out of sorts. It was an unwelcome swathe of volunteer corn plants poking their tasseled heads up in a plot where ProdiGene had planted one of its undisclosed pharmaceutical crops during the previous growing year. The plot had been taken over by a new farmer who had grown and harvested soybeans during the 2002 season. Federal guidelines had expressly directed the removal of all plant debris and seed from the fields and had forbidden the use of the same land for growing food crops for a number of years following the test.

The USDA immediately demanded that the company remove all remaining corn plants in the plot, and the department commandeered about 500 bushels of soybeans that had been harvested and transported to a storage facility in the state to keep them out of human and animal food supplies. In short order, Greenpeace scaled one of these storage silos and unfurled a banner reading "This is your food on drugs."

In the USDA press release acknowledging the incident, it was also revealed that this was not the first of ProdiGene's violations. Just a month before the Nebraska contamination was discovered, government inspectors in Iowa had found another near identical case of haphazard violation. There, volunteer biopharm corn plants that had reseeded themselves from a ProdiGene experiment the year before were found growing in a soybean field and sprouting out of piles of corn plants that the company had harvested but failed to remove from the test site as mandated in its permit.

The USDA levied fines on the company, portraying to the public a heavy upper hand in protecting food crops from such transgressions. In its initial announcement of the settlement, the department reported that ProdiGene was being forced to pay $250,000 as a penalty along with the cost of destroying the potentially tainted soybeans—an endeavor that involved not just the purchase of the beans but also their transport and burning at a hefty bill of $3.5 million. But then, in March of the following year, the *Washington Post* broke a story revealing that the government had actually subsidized the cleanup. The government had offered ProdiGene a $3.75 million no-interest loan, with a generous payback plan that let the company off the hook for a year and then gave them two more years over which to return the money—interest free—in quarterly installments. Thus, the newspaper reported, taxpayers would be subsidizing ProdiGene's errant ways.

The company had been fined and given a public hand-slapping, but not before anti-biotech advocates had put a bad taste for biopharm in the public's mouth. This contamination was a further sign, opponents said, that GM crops couldn't be trusted, especially when they carried untested drugs within their confines. A year earlier, the unapproved variety of Bt corn known as StarLink had contaminated the U.S. food supply, and now, unapproved, experimental animal drugs had been caught commingling with the potential makings of veggie burgers, tofu, and other soy-based foods. It was activist organizations' goal to ensure that these two incidents would forever be tied in the mind of the public.

The StarLink Effect on Biopharm

The StarLink case had been broken in September 2000 by a coalition of anti-biotech activist organizations known as Genetically Engineered Food Alert, which had collectively tested a variety of supermarket products for the presence of unapproved genetically modified ingredients. StarLink, a strain of corn engineered by the French biotech company Aventis Crop Sciences to express a form of the Bt *(Bacillus thuriengensis)* crystal toxin protein called Cry9C, had failed human allergenicity tests when put forward for regulatory approval in the United States. As a result, a feedlot-only sales sticker had been applied to the product, and seed companies were required to educate farmers to keep StarLink segregated and sent only to animal feed operations or ethanol fuel-producing plants.

Naively, federal regulators and the companies that cajoled them believed that this variety of corn could be grown unconfined and kept out of the human food supply. But the large-scale, industrialized food system through which commodity corn travels in the United States was never set up for such segregation. Thus, it was only to be expected that unapproved StarLink corn would soon make its debut in human food products.

Contaminated Corn

The first incident was identified in boxes of Taco Bell brand taco shells sold in a supermarket in Maryland. The taco shells were marketed by the food manufacturing giant Kraft and were a $50 million annual income generator for the company out of a total annual revenue of $27 billion. As such, it was not a matter that the company wanted to take lightly. Genetically Engineered Food Alert had issued a press release identifying the taco shells as the offender, and when Kraft's own independently commissioned tests also came back positive for the banned human foodstuff, the company decided to invoke a wide-sweeping nationwide voluntary recall of all taco shells sold under the Taco Bell brand.

In the recall, Kraft spoke out about its dissatisfaction with the current government regulatory procedures for approving genetically modified foods. All approvals of GM crops not yet cleared for human consumption should stop, the company complained, and purveyors of GM seed should also be required to provide rapid tests for detecting the presence of the GM product in food. "Kraft's suggestions have merit and deserve to be taken seriously," the head of the Biotechnology Industry Organization, Val Giddings, was quoted as saying in a *Wall Street Journal* article covering the debacle.

The same article reported that the Food and Drug Administration (FDA) did not expect any more food recalls because of StarLink. By early the following year, however, the Kellogg Company had identified StarLink in its Morningstar Farms brand of veggie burgers and meat-free corn dogs. In a matter of months, the controversy had spread as far as Bolivia and Japan. Ultimately, Aventis was forced to admit that "StarLink is forever," and that even after millions of dollars in cleanup operations requiring the rerouting of 28,135 trucks, 15,005 rail cars, and 285 barges, the nation's food supply could never be certain to be rid of the offending agent.

Not that consumers needed to worry, of course, insisted the company. In the wake of the outbreak, no confirmed cases of allergies or other public health threats had been identified. However, an optometrist in Florida named Keith Finger had testified before an Environmental Protection Agency (EPA) expert panel that upon suspicion of an allergic reaction to StarLink, he had intentionally obtained StarLink corn through the mail, confirmed its identity, and then subjected himself to a feeding experiment. The experiment sent him to the hospital suffering from itchy allergic welts and an increase in his blood pressure. But such anecdotes were merely that, insisted government regulators and company executives alike: They held no statistical significance to point a finger at StarLink. This was a publicity fiasco, not a public health disaster.

Biopharm in Peril?

While undercover activists had discovered the StarLink contamination, government regulators themselves had honed in on the transgression in the ProdiGene case as part of a routine inspection. Thus, as activists attempted to link these two events, government regulators sought to reassure the public that the ProdiGene case warranted nowhere near the level of concern that StarLink had incited. Nonetheless, the ProdiGene incident and the publicity that it generated led to a general slowdown in the biopharm race and gave even enthusiastic interest groups occasion to pause. After the fact, the USDA issued a press release confirming its commitment to strengthening field testing requirements for pharmaceutical and industrial GM crops. "These specific safeguards include comprehensive confinement procedures, performance standards, and required monitoring/auditing practices for ensuring that out-crossing or commingling with other seeds and commodities are prevented," promised the department in its press release.

Even with subsidies from the government (and the taxpayer, according to the *Washington Post*), ProdiGene's fines nearly broke the company. After all, despite a rash of field trials, ProdiGene had no products on the market and little to bring in cash other than hope and prospect. With two costly strikes against the company, investors were giving a second thought to ProdiGene's promises of profit based on PMPs. A year after headlines pasted ProdiGene's name across the nation in association with government violations in a risky new business, the company dismissed its CEO. (In 2001, at the same time that the contaminated seeds were sprouting, he had been promising that 10 percent of all corn would bear pharmaceutical genes within a decade.)

In his wake, a seasoned "turnaround" specialist, Russell Burbank, was brought in to right the ship. Burbank pledged to "work with the regulatory folks to take the public's concerns into account," and to "bring focus and discipline to the company." Soon thereafter, ProdiGene was bought by International Oilseed Distributors Inc., a subsidiary of Iowa-based Stine Seed Company, which purchased a majority ownership stake in ProdiGene in 2004 and agreed to pay off the remainder of the fledgling company's unpaid $3 million penalty levied in the contamination cases two years earlier.

It was a bad time for biotech, biopharm in particular. Following the ProdiGene fiasco, the Biotechnology Industry Organization—usually staunch supporters of any GM campaign—announced that several of its member companies were invoking a voluntary moratorium on growing biopharm crops in Corn Belt states. Food and grocery manufacturers began to add to the chorus of opposition, with organizations such as Quaker Oats and the Grocery Manufacturers of America trade group speaking out against growing experimental drugs in food crops. Then, in the fall of 2003, Monsanto pulled out of the biopharm fray, shutting down its PMP subsidiary Integrated Protein Technologies. While Burbank would receive a "Turnaround of the Year Award" in 2004 for his emergency salvage and sale of ProdiGene, the company soon disappeared off the map—its website disabled, its phone lines disconnected, its bearing as a harbinger of the biopharm revolution duly buried beneath Web searches for the company name yielding only scores of references to fines and fiascos levied years earlier. Certainly this was not the way that the biotech industry had intended things to go.

A Government Mea Culpa

The government was quick to take blame. Perhaps it was a matter of glitches and oversight in the regulatory system that had caused these failures, not the promising technology itself. In June 2004, the USDA promised a revision of its regulatory rules to allow "the public greater access to information about exactly what crops biotechnology companies are planting and how great a risk they pose." This foreshadowed a decision a couple of months later by a U.S. District judge in Hawaii that would order the USDA to publicly reveal the permitted locations of four biopharm field trials authorized in that state. But a year and a half later, little remedy had been made, and the USDA's own inspector general released a report incriminating the department and concluding that it had failed in its mission to regulate field trials of GM crops and high-risk chemical-producing crops like biopharm trials in particular.

Further condemnation came when a federal judge ruled in early 2006 that the USDA had violated federal environmental laws in permitting experimental biopharmaceutical crops to be grown in Hawaii without adequately taking into account environmental impacts. The USDA's actions had been "arbitrary and capricious" in allowing the trials to take place, said Federal District Judge Michael Seabright in his ruling. Echoing the judge's sentiment, Paul Achitoff, an attorney for Earthjustice—one of the nonprofit organizations that had brought the suit against the government—accused the USDA, saying that "the agency entrusted with protecting human health and the environment from the impacts of genetic engineering experiments" had been "asleep at the wheel."

Ventria Soldiers On

What had once been thought of as a burgeoning biopharm industry seemed headed for demise. Nonetheless, there were companies that held out, like Ventria Bioscience. Ventria's PMP rice was engineered to produce two human proteins, lactoferrin and lysozyme, the first a protein expressed in human breast milk and the latter a ubiquitous enzyme found in saliva and tears that possesses anti-microbial properties. Ventria shopped its biopharm rice trials around from state to state, starting in California, a major rice producing region in the United States, where the state's department of agriculture quickly and soundly rejected the request. Ventria next tried Missouri, where Anheuser-Busch, the state's largest rice buyer, threatened a boycott if they set down root. Then it was on to a site in North Carolina less than a mile from the rice germplasm nursery at the USDA's own Tidewater Research Station. Ventria finally settled down in Kansas, where no rice was grown and a deal was cut with a small town to bring jobs and fame to a rural locale hard hit by the farm crisis and leaning economically only on a nearby army base.

Learning from ProdiGene's mistakes, Ventria had armored itself with a privately held board that included some of the biggest names in biotech, who backed more than 85 percent of the company's financing. That economic battle front had enabled the company to weather assaults, first in its home state of California and then in Missouri, yet still push forward with its approach. Ventria had also taken pains to present its potentially controversial GM product as possessing consumer benefit or, better yet, with humanitarian implications for the developing world, hoping this warm glow could shade out naysayers and help to immunize the company against consumer revolt.

Whenever they had the chance, Ventria representatives hawked the "lifesaving" potential of rice-grown lactoferrin and lysozyme as additives to infant formula and oral rehydration solutions in the developing world, where large numbers of children die of diarrhea. While promising the "very soon" emergence of such a commercial product, which they suggested would be sold over the counter and to the U.S. Army, Ventria was also busy with a backup plan. After all, oral rehydration solution—especially when produced using transgenic GM technology that had not been cleared through clinical trials—did not promise a blockbuster market in the United States, and humanitarian applications overseas would likely be economically just that: humanitarian rather than profit-driven, if poor mothers of sick children could even be convinced to use this genetically modified miracle cure.

Thus, while Ventria was busy touting the humanitarian applications and safety and security of its outdoor-grown rice bearing human genes, the company had already published what might ultimately prove to be a more profitable application of its biopharm technology. In a research article that appeared in the *Journal of Nutrition* in 2002, scientists from Ventria and the University of California–Davis (where the company's founder was on the faculty) had described what was perhaps the true economic driver behind Ventria's biopharm rice. The article was entitled "Rice Expressing Lactoferrin and Lysozyme Has Antibiotic-Like Properties When Fed to Chicks." It began with the promise that "one strategy for replacing antibiotics in animal diets is to employ antibacterial molecules normally found along the digestive tract. Lactoferrin and lysozyme are present in mucosal secretions and in milk where they provide defense against bacteria along epithelial surfaces."

It was this potential—to grow a supplemental animal feed crop which with little processing could be delivered directly to livestock and confer antibiotic properties—that appeared more promising from a profit perspective than a GM-hydration shake produced from plants bearing human genes that had not been required to undergo any type of human safety testing.

Opposition to Biopharm Grows

Despite Ventria's persistence, sizeable opposition to biopharm crops came both from activist groups and from the technology itself. Although initially envisioned as a cheaper way to mass-produce drugs, advances in indoor fermentation technology and the perils of pharmaceutic purity requirements in the human drug realm were fast placing outdoor-grown PMPs out of the mainstream market. Making matters worse, Rick North—the indefatigable anti-rBST campaigner—was not satisfied to stick with his campaign against recombinant hormone use in milk. He had broadened his scope to include biopharm under his "Safe Food Campaign" umbrella.

In 2005, North, working through sympathetic Oregon state senators, had been instrumental in having Senate Bill 570 entered into his state's legislative agenda. The bill proposed a four-year moratorium on the growing, raising, or cultivating of genetically engineered pharmaceutical or industrial crops outdoors, or in any plant that was commonly used for human or animal food. The bill passed the State Senate but did not reach a vote in the House of Representatives before the 2005 legislative session drew to an end, thereby marking its demise. But by entering the issue into public record, North had caught the attention of the power brokers in his heavily agricultural state. By the end of the year, the Oregon governor's office and the State Senate had convened a committee charged with reviewing and developing policy recommendations for how biopharm crops might be policed in the state.

After a year of meetings and hearing testimony from biopharm experts, the committee issued a list of policy recommendations, which were signed into law by Oregon's governor on June 25, 2007. The committee recommended that potential biopharm permits in the state, of which there had been nary one since 1999, be reviewed on a case-by-case basis, but also recommended that only non-food, non-feed crops be used for outdoor growing and that any proposal for a food crop include protection by secure greenhouse production if possible. Furthermore, the policy document recommended that the Oregon establish a memorandum of understanding with the USDA that would allow information pertaining to the locale and identity of biopharm crops permitted in the state to be released to state officials (current requirements kept them hush-hush) and gave Oregon's departments of agriculture and health the authority to implement state-level reviews of any permitted applicant.

Despite the fact that Oregon had never been a big player in the biopharm race, the passage of binding biopharm regulation in a state known for being at the leading edge of environmental legislation with its historic bottle bill and forward-thinking land use policies sent the message that biopharm was no longer beneath the regulatory radar screen. The once rosy prospect of turning significant percentages of America's food crops into outdoor drug factories was beginning to pale. But biotech's field of dreams was not about to dry up, not by any means, and indeed, there was more up the GM sleeve than just drugs.

Fill 'er Up? The Biofuel Movement

With impeccable timing, just as industry enthusiasm for biopharm was beginning to wane, a new, more palatable application of GM technology in food plants arrived on the scene. Industry insiders suspected that while the public didn't want drugs in their food, they would have no problem at all with putting food in their gas tanks. Gas prices were escalating logarithmically, and even the war hawks in the Bush administration were looking for alternatives to dragging costly tanks of oil across the seas. The "biofuel economy" was making headlines as manufacturers scaled up ethanol processing plants and biodiesel refineries in anticipation of the new life biofuels were predicted to offer to the sagging farm economy. In early 2006, President Bush announced in his State of the Union address a move toward "homegrown" fuel sources and "bio-based transportation fuels." Not only would this solution draw attention away from the "war for oil" mantra that the Bush administration had been plagued with; it also promised to prop up agricultural commodity prices and make the farm belt a profitable playing field once again.

Although consumers were anticipated to have greater tolerance for genetically modified fuel crops as compared to genetic tinkering with their food, at the onset of biofuel prospecting, the two were not entirely separable. In the United States, the majority of ethanol was produced from corn, while in Europe, biodiesel produced from rapeseed (or canola) led the way to the pump. Of course, both corn and canola were members of the biotech "big four," joined by cotton and soy at the helm of the GM empire. At the very least, the potential that GM varieties of these crops held for boosting up yields—considered one of the gravest challenges in converting widespread fuel consumption from oilfields to farms—could be touted to further the GM cause. Europeans wrestled with the ethics of promoting GM rapeseed for fuel despite their distaste for GM food. In Brazil, President Luiz Inacio Lula da Silva issued a declaration that GM soy would be channeled to biofuel, while "good soya" would be retained for food.

Many were upset by this blurring of the food/fuel distinction when it came to genetically engineering in new, as yet unapproved traits designed to increase fuel potential of a food crop. Groups like Greenpeace and Friends of the Earth were rankled, and scientists were as well. In a letter to the editor in the journal *Nature Biotechnology*, pro-GM plant scientist C. Neal Stewart pointed out the danger of pursuing genetically modified food crops for biofuel: "Other than the potential ethical dilemma of having to choose whether to use biomass for fuel rather than food or feed (although the use of stover obviates that question), genetic modification of a food crop for industrial purposes poses biosafety and regulatory issues that might be irreconcilable." His letter went on to reference the Anheuser-Busch attack on Ventria as well as the Grocery Manufacturers of America's commentary to the USDA on its concern over keeping pharmaceutical or industrial transgenes out of foodstuff.

Stewart's reference to "stover"—the unprocessed husks and straw left over after harvest—was a nod to the new direction that the burgeoning biofuels industry was taking, which offered to lead the trajectory away from food crops and to generate vast new markets for genetically modified products at the same time. As of 2008, the large majority of biofuel ethanol produced globally came from corn kernels or sugarcane. This was costly, ethically debatable in its pitting of potential human food calories from these sources against the production of industrial fuels, and even in the wildest GM dreams would never come close to satisfying the world's voracious appetite for fuel.

Cellulose: A Golden Opportunity?

But the technology was in place. It involved a relatively simple series of steps in which the starches and sugars in corn or sugarcane could be fermented by yeast into ethanol, utilizing an age-old biochemical pathway more in the realm of distilleries and breweries than big biotech. But even using sweet starting materials like corn or sugarcane produced relatively inefficient yields of alcohol and left behind as a waste product a fibrous mass of cellulose (the stuff that gets stuck in the teeth when eating corn on the cob). Certainly, there had to be other sources of ethanol and other means by which to extract it. Not coincidentally, it was buried in this biomass of cellulose waste product that biofuel scientists saw a potentially golden biotech opportunity.

Cellulose is the primary component of the cell wall that encloses each cellular building block in a plant, providing rigidity and toughness to the plant's leaves, stems, and shoots. After corn is harvested, its golden cobs removed, and the field left for scarecrows, it is primarily cellulose that is left behind in the form of husks and stalks. Cellulose is found not just in corn husks but in every plant growing on earth; it is estimated that half of all organic carbon on the earth is caught up in the long strands of cellulose wrapped around cell walls of plants.

Despite its fibrous feel and tasteless flavor, cellulose is made up solely of sugars—a string of alternately oriented glucose monomers—which quite happily and enthusiastically enter the fermentation pathway and rearrange themselves to produce ethanol when treated with the appropriate enzymes, called cellulases. Although cellulose on its own stands ready and willing to enter the fermentation pathway leading to ethanol production, wresting cellulose away from the other components of the cell wall—a more complex carbohydrate called hemicellulose and an utterly inscrutable tangle of carbon biochemistry known as lignin—is a necessary step before enzymatic ethanol production can take place. This has proven challenging, requiring pressurized steam treatment and harsh acid baths that often leave inhibitory residues that interfere with subsequent enzymatic processing.

Untangling the Lignin Dilemma

It is primarily the presence of lignin that interferes with the rapid entry of plant-derived cellulose into the ethanol pathway. Lignin stands second to cellulose as a carbon source on the earth, commandeering an estimated 30 percent of all earthly organic carbon found outside of fossil fuel. Unlike cellulose, which stands in neatly repetitive crystalline chains of simple sugar building blocks, lignin is biochemically chaotic, made up of randomly repeated substructures that are often cross-linked and catalyzed into meshy polymers. While cellulose strengthens a plant's cell wall, lignin gives plant matter its woody feel, helping to hold the plant up and transport water up its height.

For a plant fighting it out to survive windstorms and reach the sun, to fend off pests and to hydrate its highest reaches, lignin is an essential component, one that would be sorely missed in any plant growing in the wild. But for a plant intended as a cultivated fuel source for hungry combustion engines, lignin is a liability, a messy barrier to extracting the maximum amount of energy in the quickest, cheapest manner. This challenge did nothing to deter biofuel scientists, though. If anything, it acted as a rallying cry for the biofuel race: Who would be first to find a solution to this tangled conundrum, a technological trick enabling industry to tap into the vast potential to transform ton upon ton of woody, lignocellulose plant matter into ethanol?

If the potential to overcome the lignin dilemma in order to more effectively use cellulose to make ethanol could be brought to the table, food plants would no longer necessarily be part of the discussion. After all, every living plant—not just those that feed us—contains cellulose. Indeed, many food plants have been bred to retain less cellulose than their wild progenitors, making wild or weedy strains more advantageous in some cases. With biofuel on the menu, the recipe suddenly changes. Now, plants with particularly pulpy propensities would be selected for and modified in a manner that would drive them toward fueling the future.

Turning away from food crops, scientists set their sights on two of the most promising biofuel crops, switchgrass and poplar. Switchgrass, or *Panicum virgatum* as it is known scientifically, is a perennial grass species native to the North American prairie. It is now well established across nearly the entire United States, save for the far western coastline states of California, Oregon, and Washington. Research on breeding and cultivation of switchgrass has been conducted by the USDA since the 1930s, but it was only in 1990 that the research focus shifted to include the biofuel potential of this prairie native.

Switchgrass has historically been planted through the Conservation Reserve Program (CRP), a USDA-funded program that pays farmers to plant land with long-term resource-conserving species in an attempt to enhance environmental quality as well as promote wildlife conservation. Switchgrass is a perennial, returning year after year, and propagates via both seeds and spreading underground rhizomes. Thus, theoretically, switchgrass stands could be harvested for biofuel each year and with proper management return to reap yields again in subsequent seasons. Certainly nobody would frown on propagating a native species for sustainable biofuel harvesting, would they?

The second plant species being proposed for cellulosic ethanol production is poplar, a woody, deciduous tree favored by the pulp and paper industry for its speedy growth, pulpy fiber, and pale color. Indeed, poplar has a long human history. Leonardo da Vinci's *Mona Lisa* was painted on poplar, and the round wooden boxes in which Camembert cheese is contained have historically been made of poplar. Today, poplar is a favored product for the snowboard industry because of its light and flexible properties.

By the time the biofuel debate rolled around, biotech scientists had already long been at work attempting to engineer a "better poplar" with increased growth and reduced lignin conent. Because of its pithy, tangled structure, lignin was also the bane of the pulp and paper industry. It was tolerated out of necessity but that was economically and environmentally costly, due to the caustic nature of the chemicals used to remove it and the scale of production. Given that wood pulp production in the United States exceeds 80 million tons each year, and that another 30 million tons of lignin are extracted in the process, the industry was hard pressed to find a solution to the so-called lignin problem.

Thus, scientists worked diligently to unravel the genetics of the lignin biosynthesis pathway and had created transgenic trees expressing less lignin in their cell walls. But such genetically engineered trees were met with fierce resistance from environmental groups and some scientists. Reducing lignin would lead to forests of "floppy trees," these groups asserted, and could potentially make trees more susceptible to pest penetration and damage as well. Because both switchgrass and poplar reproduced through pollination and were native to North America, the potential for spread of these GM varieties into wild or weedy relatives was a serious possibility.

There were other problems beyond the issue of engineering "floppy plants" that, without their lignin skeleton, might be more susceptible to wind and weather. Even once lignin had been reduced or removed in a plant, the pesky process of breaking the remaining cellulose down into its constituent glucose sugars (a necessary step before it could be fermented into ethanol) persisted. Other GM efforts were being directed at encoding bacterial or even insect-borne enzyme genes into biofuel plants, including corn as well as switchgrass and poplar, which would degrade the cellulose in the cell walls making the energy in the plant more accessible. Unless such properties could be put under regulatory control so that breakdown would begin post-harvest. some variant on the "floppy tree" hypothesis might arise, not to mention the dire consequences of feeding kernels of cellulose-degraded corn into the food supply.

To solve the lignin dilemma, scientists are now attempting to harness the once feared wood-degrading properties of termites into plants. Although when it comes to wood dwellings, we tend to think of termites as the ultimate destroyers, interestingly, it is not termites themselves that possess the ability to degrade wood. Hiding out within the hindgut of termites are an array of microbes that do the work for them. As lignocellulose from a termite's woody meal travels through the insect's digestive tract, it is degraded by resident bacteria possessing unique cellulase enzymes capable of breaking down the cellulose into its constituent glucose monomers. Some of these bacteria are species similar to those found in ruminant guts, aiding animals like sheep and cows in digesting grass and other fibrous plant material. Other enzymes, though, appear to be unique within the narrow slice of life that inhabit termite intestines. By sequencing the genes for such enzymes and encoding them into plants themselves, or at the very least bacterial factories, scientists are aiming to engineer self-digesting wood ready for the refinery. Such a far-fetched scenario sounds, blatantly and literally, like a serious case of not seeing the forest for the trees.

Biofuel and Beyond

Much of this is speculation, of course, of the kind that has been heard off and on since the nascent days of biotechnology's 1970s debut. Whether agricultural biotechnology will bring us hope, hype, or horror in the urgent quest to find new fuel sources remains to be seen. What is certain, though, is that these technologies will not take place without debate, as history has shown again and again. As agricultural biotechnology moves away from food in its foray toward biofuel, consumers may find themselves less wary of such tinkering with nature in the name of powering their vehicles and heating their homes. Or perhaps this quest may backfire, bringing greater attention and awareness to the presence of GM crops in our midst and fueling a nascent and unexpected consumer opposition. Whatever course agricultural biotechnology brings us down, consumers would do best to be informed and take a participatory rather than passive stance when it comes to the future of our food, farms, and forests.

The Ongoing Battle Over GM Food

Whence the future of GM food, in the wake of ongoing controversy and a renewed urgency over global food security? Will GM technology pull ahead, feeding the world's stomachs while filling their gas tanks? Or will repeated bouts of resistance finally bring the GM approach down, relegating it to the "been there, done that" graveyard of technologies that failed to deliver on their lofty promises?

In all likelihood, neither of these scenarios will come to pass, at least not in entirety. The stalwart GM traits that have already saturated the market through proprietary patents (such as tolerance to herbicides like Monsanto's Roundup or the presence of insecticidal toxins encoded by genes from the Bt soil bacterium) are without doubt here to stay. Adoption of such crops in the United States has continued to escalate since their introduction; in 2008, 80 percent of all corn, 86 percent of all cotton, and 92 percent of all soybeans grown in the United States were GM varieties. In the future, these already established GM traits are likely to continue their spread beyond the initial "big four" industrial crops of corn, soy, cotton, and canola to reach deeper beneath the surface of agricultural commodities and infiltrate a wider range of common produce crops. However, these forays into the food supply will not be immune to market pressure and consumer revolt.

Indeed, in 2004, Monsanto announced its plans to drop the release of its Roundup Ready wheat varieties, which it had succeeded in developing, but which had met with stiff opposition from not only activist groups but farmers and wheat industry organizations as well. The company had earlier been forced to take its NewLeaf GM potatoes off the market back in 2001, after McDonald's refused to sell French fries made from potatoes engineered to express the Bt toxin. And in 2005, Monsanto received USDA approval to market Roundup Ready sugar beet seeds, but such seeds were not sown at the time because of concerns over consumer opposition. Pressure does not appear to be letting up either. The August 2008 announcement that Monsanto planned to divest its rBST operation showed that consumers could turn the tide against a major corporate player.

On the surface this may seem like failure, but the gene giants are far too strong to just give up. In early 2008, it was announced that GM sugar beets would finally be planted in Oregon's Willamette Valley, which supplies 90 percent of the nation's beet seeds. On cue, a lawsuit was filed by a coalition of environmental groups and organic seed growers, in the same Northern California U.S. District Court that had issued an injunction blocking the sale of Roundup Ready alfalfa seeds just a year before because of environmental oversight. This tug-of-war trajectory between the development of new GM crops and consumer and legal challenges to their use is likely to continue to dog the future expansion of GM crops, but it will probably come nowhere close to shutting down the GM mission in entirety. While legal challenges and consumer bickering make their way through the courts and complicate customer acceptance, behind the scenes, the GM march continues unabated.

In 2007, the USDA approved the planting of eleven new pharmaceutical or industrial crops. They included a safflower strain developed by the company SemBioSys Genetics, engineered to produce growth hormones from carp and intended to be used in the aquaculture industry to "beef up" farmed fish. In early 2008, the USDA announced plans to allow the field release of another safflower strain developed by the same company, this time engineered to produce the human insulin precursor molecule called proinsulin within its seeds. Meanwhile, Germany approved new and more restrictive labeling laws for GM-free foods in early 2008. Despite the slow infusion of GM food into Europe forced in the wake of the EU's loss of the World Trade Organization lawsuit in 2007, consumer opposition there remains steady and staunchly resistant to GM products on supermarket shelves.

Perhaps the more pressing question that this book leaves in its wake pertains not to the nattering that has been going on for so long now between the United States and European nations over the acceptance of established GM foods in these countries, but to the future that GM food might hold for the more pressing and urgent matter of food security in the developing world. Can GM seeds solve the increasingly dire dilemma of world hunger?

Not on their own, and certainly not in their current form. Biotech's "big 5"- BASF, Monsanto, Bayer, Syngenta, Dupont and their partners- are reported to have already filed over 500 patent documents pertaining to 'climate ready' genes in food plants. As these companies gobble up exclusive rights to genetic traits essential for agricultural adaptation to weather shifts impacting the world over, those in poor countries suffering from food insecurity are unlikely to be first in line to reap the benefits.

Real progress towards shifting the balance of food insecurity in poor nations will not be made until the techniques of agricultural biotechnology are given their just public due, and cease to exist primarily as tools for corporate profit. Beyond bridging the public-private divide in research and access to agricultural biotechnology, new approaches to agricultural productivity in poor nations must take into account and involve directly the stakeholders that such approaches profess to serve. Economically profitable quick fixes and panacean sound bites may benefit big business, but will do little to improve the lives of the majority of those who go to sleep hungry each night.

It is here where one might hope that new agricultural strategies—likely those taking more balanced and sustainable approaches based on conventional agricultural breeding and management practices—might be able to draw upon GM technologies more as a condiment than as a main course, served up in the name of true public good rather than corporate profit. In this context, one might hope that new tactics will be developed that use such laboratory tools to fine-tune and evaluate the efficacy of methods truly intended, from start to finish, to address the real-life food struggles that lie at the intersection of food, politics, and technology. It is here, at this most critical of crossroads, that the sagas in this book leave off, encouraging us to learn from the grass-roots level of those living such stories and to pledge not to let politics and power stand in the way of solutions that utilize scientific knowledge and technologies in their most altruistic manner possible.

Notes

Preface

the most recent . . . . : The Mellman Group, *Public Sentiment About Genetically Modified Food (2006)*, Pew Initiative on Food and Biotechnology, Pew Charitable Trusts, 2006.

Chapter 1

"U.S. Ready to . . ." : Edward Alden, "U.S. Ready to Declare War Over GM Food," *Financial Times*, January 9, 2003.

"There is no point in testing . . ." : Elizabeth Becker, "U.S. Delays Suing Europe Over Ban on Modified Food," *New York Times*, February 5, 2003.

"Because of Cubbie's support for . . ." : "Fried Politics: Restaurant serves 'freedom fries,' " CNN.com, February 19, 2003.

officially took the French out . . . : Sheryl Gay Stolberg, "An Order of Fries, Please, But Do Hold the French," *New York Times*, March 12, 2003.

"I have no animosity toward . . ." : Andrew Jacobs, "Strained Relations, And Spilled Merlot; Restaurateur Shows Displeaseur Over French Leaders' Stance on Iraq," *New York Times*, March 8, 2003.

The official press release documenting . . . : "U.S. and Cooperating Countries File WTO Case Against EU Moratorium on Biotech Foods and Crops: EU's Illegal, Non-Science-Based Moratorium Harmful to Agriculture and the Developing World," press release, Office of the U.S. Trade Representative, U.S. Department of Agriculture, May 13, 2003.

"many Europeans must agree . . ." : Transcript released by U.S. Department of Agriculture, Press Conference with Agriculture Secretary Ann M. Veneman and U.S. Special Trade Representative, Ambassador Robert T. Zoellick regarding the EU Moratorium on Biotech Crops and Food, May 13, 2003.
polls indicated that . . . : Pew Initiative on Food and Biotechnology, "Public Sentiment About Genetically Modified Food," March 2001.

the vehement debates . . . : For an overview and documentation of the debates surrounding the birth of recombinant DNA technology in the 1970s, see James D. Watson and John Tooze, *The DNA Story: A Documentary History of Gene Cloning* (W. H. Freeman & Co., 1983).

"new kinds of hybrid . . ." : James D. Watson and John Tooze, *The DNA Story* (W. H. Freeman & Co., 1981), page 5.

"marvelous achievement of science . . ." : Ibid, page 3.

"touch on ethical issues . . ." : Ibid, page 42.

"this is like asking . . ." : Ibid, page 49.

discovered by a local tabloid . . . : Arthur Lubow, "Playing God with DNA," *New Times*, January 7, 1977. This article provides a colorful background to the local DNA debates taking place in Cambridge at the time.

"the occasional rare exposure . . ." : Watson and Tooze, page 117.

"All of the nations . . ." : Ibid.

Past threats to . . . : Ibid.

"Biotech's first superstar" . . . : Joan Hamilton, "Biotech's First Superstar," *Business Week*, April 14, 1986.

"Shaping Life in the Lab" . . . : Frederic Golden, "Shaping Life in the Lab," *Time*, March 9, 1981.

Chapter 2

And so it makes sense . . . : For an overview of the discovery of the Roundup Ready genes by Monsanto, see Daniel Charles, *Lords of the Harvest: Biotech, Big Money, and the Future of Food* (Perseus Books, 2002).

Since the company was not itself . . . : In 1992 and again in 1993, Monsanto signed licensing agreements for its Roundup Ready soybeans with Pioneer Hi-Bred International, not fully anticipating the economic return the patented varieties would bring. Later, evidence of price-fixing and lawsuits between the two companies would arise. These behind-the-scenes negotiations became public in 2004. See David Barboza, "Questions Seen on Seed Prices Set in the 90s," *New York Times*, January 6, 2004.

wrote effusively in praise . . . : Rachel Carson, *Silent Spring* (Houghton Mifflin, 1962), page 289.

"the main technical problem . . ."* : Ibid, page 290.

The Flavr Savr tomato was a perhaps unlikely victor . . . : For an overview of the development of the Flavr Savr tomato, see Belinda Martineau, *First Fruit: The Creation of the Flavr Savr Tomato and the Birth of Biotech Foods* (McGraw-Hill, 2001).

More than 10,000 replies . . . : "UK Prince sparks GM food row," *BBC News*, June 1, 1999.

in an article published . . . : HRH Prince Charles, "My 10 fears for GM food," *Daily Mail*, June 1, 1999.

"Combine lots of . . ." : Online, January 13, 2000, available online at www .reason.com/bi/bi-gmf.html. Accessed February 18, 2008.

first emerged in an . . . : Molly O'Neil, "Geneticists' Latest Discovery: Public Fear of 'Frankenfood,'" *New York Times*, June 28, 1992.

Accompanying a September 1976 . . . : George Wald, "The Case Against Genetic Engineering," *The Sciences*, September 1976, page 6.

The following year . . . : James Watson, "An Imaginary Monster," *Bulletin of the Atomic Scientists*, May 1977, page 12.

Even Rolling Stone *magazine* . . . : Michael Rogers, "The Pandora's Box Congress," *Rolling Stone*, June 19, 1975, page 37.

Chapter 3

this wide-sweeping initiative proposed . . . : Gottfried Schatz, "The Swiss Vote on Gene Technology," *Science*, September 18, 1998, page 1810.

"Acceptance of the initiative . . ." : Ibid.

larger than life-size . . . : Personal interview with a Greenpeace campaigner, May 30, 2003.

stemming out of the . . . : For an overview of the UN Conference on Environment and Development held in Rio on 1992, see the official United Nations website at www.un.org/geninfo/bp/enviro.html.

a short outline of . . . : Rio Declaration on Environment and Development, United Nations publication, Sales No. E.73.ii.A.14 and corrigendum. Available online at www.unep.org/Documents.Multilingual/Default.asp?DocumentID=78&ArticleID=1163.

"Crossing the street . . ." : Interview with a scientist, June 4, 2003.

met six times . . . : An excellent overview and history of the negotiations leading to the Cartagena Protocol can be found in "The Cartagena Protocol on Biosafety: A Record of the Negotiations," Secretariat of the Convention on Biological Diversity, September 2003. Available online at www.cbd.int/doc/publications/bs-brochure-03-en.pdf.

At 3:30 A.M. . . . : Report of the Sixth Session of the Open-Ended Ad Hoc Working Group on Biosafety and the First Extraordinary Session of the CBD Conference of the Parties: 14–23 February 1999, International Institute for Sustainable Development (IISD), Volume 9, Number 117, February 26, 1999. Available online at www.iisd.ca/vol09/enb09117e.html.

But some victories . . . : Chee Yoke Ling, "U.S. behind collapse of Cartagena biosafety talks," Third World Network. Available online at www.twnside .org.sg/title/cheey-cn.htm. Accessed February 19, 2008.

"it was important . . ." : Ibid.

"international interests and . . ." : Ibid.

"history will not . . ." : Ibid.

"This Protocol shall . . ." : *The Cartagena Protocol on Biosafety: Text of the Protocol,* Convention on Biological Diversity, available online at www.cb d.int/biosafety/protocol.shtml.

In 1998, the small . . . : Nicholas Kalaitzandonakes and Jos Bijman, "Who is driving biotechnology acceptance?" *Nature Biotechnology,* April 2003, page 366.

according to an article . . . : Ibid.

an article entitled . . . : Nicholas Kalaitzandonakes, "Cartagena Protocol: A New Trade Barrier?" *Regulation,* Summer 2006, page 18.

they staged vociferous protests . . . : Personal interview with a Greenpeace campaigner, May 30, 2003.

what is known as the "cognitive miser" . . . : S. T. Fiske and S. E. Taylor, *Social Cognition*, 2nd ed. (McGraw-Hill, 1991).

a European survey in which . . . : George Gaskell et al., "Biotechnology and the European Public," *Nature Biotechnology*, September 2000, page 935.

studies of media coverage of biotechnology . . . : George Gaskell and Martin Bauer, eds., *Biotechnology 1996–2000: The Years of Controversy* (National Museum of Science and Industry, 2001).

the largest per capita number . . . : Nick Valery and Laza Kekic, "Innovation: Transforming the way business creates," white paper of *The Economist* Intelligence Unit, May 2007.

highest per capita consumption . . . : *Organic Farming in Europe 2005: Market, Production, Policy and Research* (FIBL—Forschunginstitut fuer biologischen Landbau, Frick, Switzerland).

In 2004, 44 percent . . . : Swiss Federal Office for Statistics, BFS.

"Not that emotions . . ." : Personal interview with a biochemist, May 26, 2003.

"Thinking from the belly . . ." : Personal interview with a scientist, June 2, 2003.

"turned into a sect . . ." : Personal interview with a neuroscientist, June 4, 2003.

Mary Shelley's eighteenth-century . . . : Mary Shelley, *Frankenstein: or, The Modern Prometheus*, 3rd ed. (Lackington, Hughes, Harding, Mavor and Jones, 1818).

"When I saw those . . ." : Personal interview with a Greenpeace campaigner, May 30, 2003.

"simply trying to fool . . ." : Personal interview with a molecular biologist, June 4, 2003.

An American meal . . . : Jaclyn Maurer Abbot and Carol Byrd-Bredbenner, "The State of the American Diet: How Can We Cope?" *Topics in Clinical Nutrition*, July/September 2007, page 202.

Consumers shelled out . . . : Eric Schlosser, *Fast Food Nation* (Houghton Mifflin, 2001), page 3.

By the end of the twentieth century . . . : Birgit Meade and Stacey Rosen, "Income and Diet Differences Greatly Affect Food Spending Around the Globe," *Food Review*, September-December 1996, page 39.

Americans spent roughly . . . : Maurer Abbot and Byrd-Bredbenner.

A steak in Denmark . . . : Lorraine Mitchell, "U.S.-EU Food and Agriculture Comparisons," USDA Economic Research Service, publication WRS-04–04, 2004.

"If Americans are willing . . ." : Personal interview with an anti-GM activist, May 30, 2003.

StarLink was a genetically . . . : Michael R. Taylor and Jody S. Tick, *The Star-Link Case: Issues for the Future* (Pew Initiative on Food and Biotechnology, 2001).

a random check of . . . : Marc Kaufman, "Biotech critics cite unapproved corn in taco shells," *Washington Post*, September 18, 2000.

Attempting to avert . . . : Statement by the U.S. Department of Agriculture and the Environmental Protection Agency, Release Number 0345.00, September 29, 2000.

In the meantime . . . : Andrew Pollack, "No Altered Corn Found in Allergy Samples," *New York Times*, July 11, 2001.

they spouted statistics of . . . : David Leonhardt, "Talks Collapse on U.S. Efforts to Open Europe to Biotech Food," *New York Times*, June 20, 2003.

what President George W. Bush called . . . : "US in new global push," *BBC News*, June 23, 2003.

in a speech before . . . : "Remarks by the President at the BIO 2003 Convention Center and Exhibition," *PR Newswire*, June 23, 2003.
"If you're going to . . ." : Edward Alden, "US retaliation against Egypt hits trade plans," *Financial Times*, July 6, 2003.

the WTO investigation snaked . . . : "GM Crops and Food, WTO dispute, Detailed timeline." GeneWatch UK, available online at www.genewatch.org/sub-555283. Accessed February 19, 2008.

petitions signed by . . . : Ibid.

the final report . . . : "Reports out on biotech disputes, World Trade Organization Dispute Settlement," September 29, 2006. Available online at www.wto.org/english/news_e/news06_e/291r_e.htm.

one of the first countries . . . : Daniel Wueger, "Consumer Information on GM-Food in Switzerland and WTO Law," NCCR Trade Regulation Working Paper No. 2/2006, September 2006.

Chapter 4

In the ad . . . : Copies of the advertisement are available on the website "Landscapes of Global Capital: Portraits of Third World Poverty," at .stlawu.edu/~global/pagescapital/poorportraits.html. Accessed February 19, 2008.

reappeared again that year . . . : Ibid.

Monsanto's website . . . : Video clips of farmers from developing countries can be viewed on the Monsanto website at www.monsanto.com/biotech-gmo/asp/country.asp. Accessed February 19, 2008.

these same videos . . . : A copy of the Monsanto video of Thandiwe Myeni can be found on the youtube website at www.youtube.com/watch?v=4suRL k8OwGI&feature=related. Accessed February 19, 2008.

a brief endorsement . . . : Monsanto's video clip of Dr. Norman Borlaug can be found at www.monsanto.com/biotech-gmo/asp/experts.asp?id=Nor manBorlaug. Accessed February 19, 2008.

"forgotten benefactor of humanity" . . . : Gregg Easterbrook, "Forgotten Bene-factor of Humanity," *Atlantic Monthly*, January 1997, page 75.

the story of an agricultural . . . : For a detailed overview of Norman Borlaug and the history of the Green Revolution, see Leon Hesser, *The Man Who Fed the World: Nobel Peace Prize Laureate Norman Borlaug and His Battle to End World Hunger* (Durban House, 2006).

In 1944, Mexico . . . : Kathleen Phillips, " 'Father of the Green Revolution' receives National Medal of Science," Texas A&M University press release, March 2006.

Farmer compliance was high . . . : Norman Borlaug, "The Green Revolution: Peace and Humanity," *Atlantic Monthly*, January 1997.

By the 1950s . . . : Phillips.

a legacy of colonialism . . . : For an overview of the Green Revolution in India, see Govindan Parayil, "The Green Revolution in India: A Case Study of Technological Change," *Technology and Culture*, October 1992, page 737.

the doomsday tome . . . : William and Paul Paddock, *Famine—1975! America's Decision: Who Will Survive?* (Little Brown & Company, 1967).

when Borlaug came knocking . . . : Parayil.

In 1966, under . . . : *Pursuit and Promotion of Science—the Indian Experience*, Chapter 21 (Indian National Science Academy, 2001).

India had already begun . . . : Vandana Shiva, "The Green Revolution in the Punjab," *The Ecologist*, March-April 1991.

overall global performance . . . : Tom Philpott, "The Revolution Will Be Criticized," *Grist Magazine*, September 27, 2006; R. E. Evenson and D. Gollin, "Assessing the Impact of the Green Revolution, 1960–2000," *Science*, May 2, 2003, page 758.

Borlaug had begun . . . : Hesser.

"just one organization . . ." : *Monsanto 2006 Pledge Report: The Sum of our Commitments*, page 4.

a product emerged . . . : For a concise review of the science and development of Golden Rice, see Salim Al-Babili and Peter Beyer, "Golden Rice—five years on the road—five years to go?" *Trends in Plant Science*, December 2005, page 1360.

According to the . . . : "Micronutrient deficiencies," World Health Organization Fact Sheet. Available online at www.who.int/nutrition/topics/vad/en/. Accessed February 18, 2008.

In India, consumption . . . : " 'Golden Rice' and Vitamin A Deficiency Fact Sheet," Friends of the Earth Safe Food Campaign. Available online at www.foe.org/safefood/rice.html. Accessed February 18, 2008.

Potrykus had been mesmerized . . . : The story of Ingo Potrykus's development of Golden Rice is recounted from a personal interview, June 5, 2003; and from Ingo Potrykus, "The 'Golden Rice' Tale," available online at www.agbioview.listbot.com. Accessed February 18, 2008.

fourteen children died . . . : Umesh Kapil, "Update on vitamin A-related deaths in Assam, India," *American Journal of Clinical Nutrition*, October 2004, page 1082.

Potrykus farmed this . . . : Potrykus.

"natural form of . . ." : "A Natural Form of Genetic Engineering," *Pathbreakers* (University of Washington Office of Research, November 1996); see also Mary-Dell Chilton, "Agrobacterium, A Memoir," *Plant Physiology*, January 2001, page 9.

produced only around . . . : Al-Babili and Beyer.

"I am happy to . . ." : Ingo Potrykus, "Genetically Engineered 'Golden Rice' Is Fool's Gold: Response from Prof. Ingo Potrykus," February 11, 2001, available online at www.biotech-info.net/IP_response.html. Accessed February 18, 2008.

"What these radical . . ." : "Potrykus Rebuts Attack on Golden Rice, " *AgBioWorld Newsletter*, June 27, 2000, available online at www.agbioworld.org/newsletter_wm/index.php?caseid = archive&newsid = 397. Accessed February 18, 2008.

Scruggs was one . . . : Adam Liptak, "Saving Seeds Subjects Farmers to Suits Over Patent," *New York Times*, November 2, 2003.

between 1994 and 2007 . . . : Ned Stafford, "GM patent rejected after 13 years," *Nature*, May 4, 2007.

The result was shocking . . . : Potrykus, "The 'Golden Rice' Tale."

"It seemed to me . . ." : Ibid.

"Life is not an . . ." : "Say no to genetic engineering," Greenpeace International campaign, available online at www.greenpeace.org/international/campaigns/genetic-engineering. Accessed February 18, 2008.

"could forever destroy . . ." : Vandana Shiva, "The Indian Seed Act and Patent Act: Sowing The Seeds of Dictatorship," *Zmag*, February 14, 2005.

"a bit of further thinking . . ." : Potrykus, "The 'Golden Rice' Tale."

Chapter 5

CNN's MoneyLine *newscast* . . . : "Stocks Down Again Amidst Uncertainty of Possible War," CNN Lou Dobbs MoneyLine, January 21, 2003, 18:00 ET. Transcript available online at transcripts.cnn.com/TRANSCRIPTS/0301/21/ mlld.00.html.

"They can play . . ." : Andrew Bolt, "A bitter harvest," *The Herald Sun* (Australia), June 8, 2007; P. Martin and N. Itano, "Greens accused of helping Africans starve," *Washington Times*, August 30, 2002.

Ten days earlier . . . : David Teather, "US Trade war threat as Europe bars GM crops," *The Guardian*, January 10, 2003.

In a commencement address . . . : "President Delivers Commencement Address at Coast Guard," press release, Office of the Press Secretary, The White House, May 21, 2003.

"Eat this or die . . ." : "Eat this or die: The poison politics of food aid," Greenpeace International, September 30, 2002, available online at www .greenpeace.org/international/news/eat-this-or-die. Accessed February 18, 2008.

"lay the basis for . . ." : "The History of America's Food Aid," Fifty Years of Food For Peace Conference, USAID, available online at www.usaid.gov/ our_work/humanitarian_assistance/ffp/50th/history.html. Accessed February 18, 2008.

"a workable scheme . . ." : World Food Programme History timeline, September 1960, available online at www.wfp.org/aboutwfp/history/instit_time line.asp?section = 1&sub_section = 1#. Accessed February 18, 2008.

"Food is strength . . ." : "The History of America's Food Aid."

"Small nations like . . ." : Personal interview with Honorable Abel Chambeshi, Minister of Science and Technology, Zambia, August 5, 2003.

"The principal beneficiary . . ." : Quoted in Dale Jamieson, " Duties to the Distant: Aid, Assistance, and Intervention in the Developing World," in *Current Debates in Global Justice*, edited by Gillian Brock and Darrel Moellendorf (Springer, 2005), page 159. This statement has since been removed from the USAID website.

In a press release . . . : "Collaborative Agricultural Bitoechnology Initiative: Mobilizing New Science and Technology to Reduce Poverty and Hunger," U.S. Agency for International Development press release, June 12, 2002.

snubbed its nose at . . . : "Charity rejects US food aid gift," *BBC News*, August 17, 2007.

then pays to have . . . : "Cargo preference requirements: Their impact on U.S.food aid programs and the U.S. merchant marine," U.S. General Accounting Office Report to the Chairman, Committee on Agriculture, House of Representatives, June 1990.

a U.S. Government Accountability Office study . . . : "Foreign Assistance: Various Challenges Impede the Efficiency and Effectiveness of U.S. Food Aid," U.S. General Accounting Office Report to the Committee on Agriculture, Nutrition, and Forestry, U.S. Senate, April 2007.

that derives 90 percent of . . . : Brooke Williams, "Windfalls of War," Center for Public Integrity profile, Chemonics International Inc., available online at www.publicintegrity.org/wow/bio.aspx?act = pro&ddlC = 8. Accessed February 18, 2008.

"Chemonics maintains a staff . . ." : Africa Dissemination Service, FEWS NET, available online at earlywarning.usgs.gov/adds/overview.php. Accessed February 18, 2008.

"Those aren't starving Zambians . . ." : Personal interview with the relative of a Zambian minister, Lusaka, Zambia, August 8, 2003.

"It's a shame they . . ." : Personal interview with a Zambian government health worker, Lusaka, Zambia, August 8, 2003.

"In the issue of...": Personal interview with a professor of agriculture at the University of Zambia, Lusaka, Zambia, August 11, 2003.

makes up 60 percent ... : Zambia Food Balance Sheet, FAOSTAT, Rome, 2002.

the history of how ... : Steven Haggblade and Ballard Zulu, "The Recent Cassava Surge in Zambia and Malawi," Conference Paper No. 11-b, New Partnership for African Development, November 2004.

Maize was brought ... : James C. McCann, *Maize and Grace: Africa's Encounter with a New World Crop* (Harvard University Press, 2005).

the earliest documented ... : D. Dodge, "Agricultural policy and performance in Zambia—History, prospects, and proposal for change." Research Series 32, University of California Institute of International Studies, Berkeley, CA, 1976, cited in Shubh K. Kumar, "Adoption of Hybrid Maize in Zambia: Effects on Gender Roles, Food Consumption and Nutrition," International Food Policy Research Institute, Report Number 100, 1994.

as some historians ... : Haggblade and Zulu.

a survey of 125 farms ... : Steven Haggblade and Gelson Tembo, "Conservation Farming in Zambia," EPTD discussion papers 108, International Food Policy Research Institute, 2003.

"I know of small-scale ..." : Personal interview with Honorable Mundia Sikitana, Minister of Agriculture, Zambia, August 7, 2003.

"Old farmers volunteered ..." : Haggblade and Zulu.

"We need to look at ..." : Personal interview with a professor, Lusaka, Zambia, August 11, 2003.

"Our relationship to the ..." : Personal interview with a government official, Zambia, August 5, 2003.

"I trooped off to . . ." : Personal interview with a pro-GM health scientist, Lusaka, Zambia, August 12, 2003.

"We were told there . . ." : Personal interview with a healthcare worker, Lusaka, Zambia, August 8, 2003.

Harvests had doubled . . . : Zambia Annual Harvest Assessment Report, FEWS NET, October 2004.

"The dynamics of . . ." : Amartya Sen, "The Possibility of Social Change," Nobel Lecture, published in *American Economics Review*, July 1999.

Chapter 6

Shiva's first notable . . . : Vandana Shiva, *Staying Alive: Women, Ecology and Development* (Zed Books, 1989).

in a book of . . . : Vandana Shiva, *The Violence of the Green Revolution: Third World Agriculture, Ecology and Politics* (Zed Books, 1992).

More than 70,000 . . . : "About Navdanya," available online at www.navdanya .org/about/organisation.htm.

the lives of India's . . . : Statistics on the agricultural workforce and income can be found at the website indiastat.com at www.indiastat.com/india/ ShowData.asp?secid = 41&ptid = 2&level = 2. Accessed February 18, 2008.

"the spread of GM . . ." : Vandana Shiva, "Food Democracy v. Food Dictatorship," *Zmag*, April 2003.

"GM rice promoters . . ." : Paul Brown, "GM rice promoters 'have gone too far,' " *The Guardian*, February 10, 2001.

"Unfortunately, Vitamin A . . . : Vandana Shiva, "The Golden Rice Hoax: When Public Relations Replaces Science," Research Foundation for Science, Technology and Ecology (Dehra Dun, India).

in a letter to . . . : Letter from Gordon Conway to Dr. Doug Parr, Greenpeace, dated January 22, 2001, available online at www.biotech-info.net/conway _greenpeace.pdf. Accessed February 18, 2008.

"Finally, I agree with . . .*"* : Ibid.

would feature her . . . : Meenakshi Ganguly, "Seeds of Self-Reliance," *Time,* August 26, 2002.

A report from . . . : Marc Morano, "Green Activist Accused of Promoting Famine Wins Time Magazine Honor," *CNSNews.com,* September 17, 2002.

"to work for . . .*"* : Gene Campaign mission statement, available online at www.genecampaign.org/ABOUT%20US/about_us.htm. Accessed February 18, 2008.

"to provide and promote . . .*"* : UPOV mission statement, available online at www.upov.int/index_en.html. Accessed February 18, 2008.

"The Indian law was . . .*"* : "Kanaga Raja, NGOs urge India to take lead, reject UPOV, Third World Network," available online at www.twnside .org.sg/title/twe294e.htm. Accessed February 18, 2008.

"save, use, sow . . .*"* : Suman Sahai, "India's Plant Variety Protection and Farmers' Rights Act," *Bridges Comment,* available online at www.icstd .org. Accessed February 18, 2008.

a two-day national symposium . . . : Suman Sahai, "GM, agriculture, and food security," *India Together,* February 21, 2008.

another event that Sahai . . . : Suman Sahai, "Monsanto Tried to Disrupt our Meeting," Letter to AgBioIndia from Gene Campaign, April 17, 2003.

approved Bt cotton . . . : "Chronology of Bt Cotton in India," India Resource Center, available online at www.indiaresource.org/issues/agbiotech/2003/ chronologyofbt.html. Accessed February 18, 2008.

Later scientific studies . . . : Shai Morin et al., "Three cadherin alleles associated with resistance to *bacillus thurengensis* in pink bollworm," *PNAS*, 2003, page 5004.

losing an average of . . . : Suman Sahai, "Bt Cotton Is a Failure," Gene Campaign, 2003.

When asked, some . . . : Suman Sahai, "Monsanto Tried to Disrupt Our Meeting."

a bigger and better meeting . . . : Suman Sahai, "Recommendations from a National Symposium on the 'Relevance of GM Technology to Indian Agricultural and Food Security,' " Gene Campaign, 2003.

received the Norman Borlaug . . . : "Subba Rao, Suman Sahai get Borlaug award," *The Hindu*, January 31, 2004.

a book entitled . . . : Devinder Sharma, *GATT and India: The Politics of Agriculture* (South Asia Books, 1994).

in a second edition . . . : Devinder Sharma, *GATT to WTO: Seeds of Despair* (Konark Publishers, 1995).

Sharma has not always . . . : Personal interview with Devinder Sharma, New Delhi, India.

"the basic idea . . ." : Nic Paget-Clarke, "Interview with Devinder Sharma: The politics of food and agriculture," *In Motion Magazine*, August 25, 2003.

"We apply innovation . . ." : This statement can be found on the home page of Monsanto's website at www.monsanto.com. Accessed February 18, 2008.

Arguing that hunger . . . : Amartya Sen, "Hunger in the Contemporary World," Discussion paper DEDPS/8, November 1997.

Sharma cited statistics . . . : Paget-Clarke.

Between 1997 and 2005 . . . : P. Sainath, "Farm suicides rising, most intense in 4 states," *The Hindu*, November 12, 2007.

a typical farmer using . . . : "The Dying Fields," *Wide Angle,* Public Broadcasting Service (PBS), August 28, 2007.

"the only viable path . . ." : Devinder Sharma, "Farmer's Suicides," *Znet,* January 24, 2004.

for the 557 million farmers . . . : Paget-Clarke.

"Next season's crop . . ." : Personal interview with Anupama (woman living in Pashtapur), Medak District, Andhra Pradesh, India, September 12, 2004.

"Seeds are our wages" . . . : Quotes in this and the following paragraphs are derived from personal interviews and discussion groups with the women of the DDS *sangham,* Zaheerabad, Andhra Pradesh, India, September 12, 2004.

"Why are Warangal . . ." : "Why are Warangal Farmers Angry with Bt Cotton?" produced by AP Coalition in Defense of Diversity and Deccan Development Society, available from Deccan Development Society, #101 Kishan Residency, Road No. 5, Begumpet, Hyderabad 500 016, Andhra Pradesh, India.

a market-research agency . . . : Mark Pearson, " 'Science,' representation and resistance: the Bt cotton debate in Andhra Pradesh, India," *The Geographical Journal,* November 29, 2006, page 306.

Chapter 7

Europe's stonewalling was . . . : Paul Meller, "Europe Rejects Looser Labels for Genetically Altered Food," *New York Times,* September 9, 2004.

On October 20, 2003 . . . : "President Bush Announces United States Intends to Negotiate a Free Trade Agreement with Thailand," press release, Office of the Press Secretary, The White House, October 20, 2003.

"Since the U.S. would . . ." : "BoT backs drive for trade pact with US," *The Nation,* March 28, 2004.

Of the two . . . : Wichit Chantanusornsiri, "US may bring up GMO issue," *Bangkok Post*, June 9, 2004.

On July 27, 2004 . . . : "Greenpeace raids Khon Kaen lab," *The Nation*, July 28, 2004.

the station had supplied . . . : "Agriculture Ministry confirms contamination," *Bangkok Post*, September 21, 2004; see also Piyaporn Wongruang, "Test seems to show GM papaya rampant," *Bangkok Post*, June 2, 2005.

The scientists and their . . . : "The great GM quandary," *The Nation*, October 5, 2004.

Rumors of whether . . . : "GMO concerns: Europeans shun Thai papaya," *The Nation*, September 3, 2004.

on August 20, 2004 . . . : "Policy reversal: Green light for GMOs," *The Nation*, August 21, 2004.

farmers and activists noisily . . . : "Genetically modified crops: Farmers, activists to take to the streets," *The Nation*, August 24, 2004.

"None of our customers . . ." : "Alarmed rice exporters join anti-GMO move," *Bangkok Post*, August 24, 2004.

had obtained a copy . . . : "Genetically modified crops."

"The government won't . . ." : Shawn W. Crispin, "Thais chew over biotech food," *Wall Street Journal*, October 29, 2004.

Before the week was . . . : "The GMO debate: Monsanto looks for a foothold," *The Nation*, August 26, 2004.

Monsanto was no newcomer . . . : "Bt cotton . . . Through the back door," *Seedling*, GRAIN, December 2001.

identified forty cases of . . . : "Agriculture ministry confirms contamination," *Bangkok Post*, September 21, 2004.

By the time all . . . : Piyaporn Wongruang, "Test seems to show GM papaya rampant," *Bangkok Post*, June 2, 2005.

an official eradication . . . : Kultida Samabuddhi, "Clean-up operation likely for GM papaya," *Bangkok Post*, September 14, 2004.

1,000 papaya trees . . . : "1000 'GM papaya trees' felled," *The Nation*, September 16, 2004.

the number one world . . . : Rudy Kortbech-Olesen, "World trade in processed tropical fruits," UN Conference on Trade and Development, June 1997.

first been identified . . . : For a history of papaya ringspot virus in Hawaii, see Dennis Gonsalves, "Transgenic Papaya in Hawaii and Beyond," *AgBio-Forum*, 2004.

another researcher by the . . . : The history of papaya ringspot virus in Thailand is taken from S. Sakuanrungsirikul et al., "Update on the development of virus resistant papaya: For the rural communities in Thailand," Proceedings of the Symposium and Workshop on Biotechnology-Derived Nutritious Foods, Bali Hilton International, February 29–March 1, 2004; and V. Prasartsee, et al., "Development of Papaya Lines that are tolerant to papaya ringspot virus," Proceedings of the JIRCAS-ITCAD Seminar The New Technologies for the Development of Sustainable Farming in Northeast, March 24, 1998, International Training Center for Agricultural Development, Khon Kaen, Thailand.

Between 1992 and 1998 . . . : Dennis Gonsalves.

the rapid rate . . . : Carol Gonsalves, David R. Lee, and Dennis Gonsalves, "Transgenic Virus-Resistant Papaya: The Hawaiian 'Rainbow' was Rapidly Adopted by Farmers and is of Major Importance in Hawaii Today," APSnet, American Phytopathological Society, August-September 2004.

A study carried out . . . : Ibid.

Researchers from countries . . . : Sarah Nell Davidson, "The Genetically Modified (GM) PRSV Resistant Papaya in Thailand: A Case Study for the Agricultural Biotechnology Policy Development in the GMS Sub-region," prepared for The Asian Development Bank and the Governments of Cambodia, Lao People's Democratic Republic (Lao PDR), Myanmar, Thailand, Viet Nam, and Yunnan Province and Guangxi Zhuang Auronomous Region of the People's Republic of China (PRC) in association with Agrifood Consulting International, Inc, USA and ANZDEC Limited, New Zealand, June 2006.

Two Thai scientists . . . : Sakuanrungsirikul et al.

a memorandum of understanding . . . : Davidson.

a minimum of fifteen proprietary . . . : Ibid.

The results were astounding . . . : Sakuanrungsirikul et al.

three parallel approaches . . . : Vasana Chinvarakorn, "Fiddling with nature," *Bangkok Post*, September 14, 2006.

Supat often spoke . . . : Personal interview with Supat Attathom, Bangkok, Thailand, February 21, 2005.

the top supplier of . . . : "Biggest exporter eyes rice renaissance," *The Nation*, June 7, 2005.

"I don't want . . ." : Personal interview with Supat Attathom.

traces of unapproved . . . : "Presence of unapproved GM rice in Europe alarms food industry-report," *Forbes*, September 12, 2006; Emma Marris, "Escaped Chinese GM rice reaches Europe," *Nature*, September 5, 2006.

returned repeated requests . . . : "China shelves commercial production of GM rice again," *Xinhua News Service*, February 26, 2007.

A minor fiasco . . . : "Regulatory Affairs—Agricultural and Environmental," *Biotechnology Law Report*, February 1, 2007, page 20.

quickly and quietly dismissed . . . : Preeyanat Phanayanggoor, Pradit Ruengdit, and Anucha Charoenpo, "Backdown on GMO farming," *Bangkok Post*, September 1, 2004.

voiced strong opposition . . . : "Alarmed rice exporters join anti-GMO move."

had been acquitted . . . : "Court clears activists of theft and trespass," *Bangkok Post*, September 16, 2006.

Chapter 8

Two books provide invaluable details and documents relating to the debates over the use of recombinant bovine somatatropin (rBST) in milk. They are Lisa Nicole Mills, *Science and Social Context: The Regulation of Recombinant Bovine Growth Hormone in North America* (McGill-Queen's University Press, 2002) and Samuel S. Epstein, *What's in Your Milk?* (Trafford Publishing, 2006).

sales figures reported . . . : Bill Richards, "Sour Reception Greets Milk Hormone—Mixing Food, Biotechnology Faces Resistance," *Wall Street Journal*, September 15, 1989.

increases of 25 . . . : Ibid; "Monsanto Told to Halt Promotion of Its Gene-Engineered Milk Drug," *New York Times*, February 13, 1991.

steadily risen each year . . . : Don P. Blayney, "The Changing Landscape of U.S. Milk Production," Statistical Bulletin Number 978, U.S. Department of Agriculture, June 2002.

a news article . . . : Debora MacKenzie, "Doubts over animal health delay milk hormone," *New Scientist*, January 18, 1992.

passed on critical data . . . : Ibid.

a report by . . . : Andrew Christiansen, "Recombinant Bovine Growth Hormone: Alarming Tests, Unfounded Approval The Story Behind the Rush To

Bring rBGH to Market," report commissioned for Rural Vermont, a project of the Rural Education Action Project (REAP), July 1995.

detailed analysis of . . . : Ibid.

in an article . . . : A. N. Pell et al., "Effects of a Prolonged-Release Formulation of Sometribove (n-Methionyl Bovine Somatotropin) on Jersey Cows," *Journal of Dairy Science*, 1992, page 3416.

claimed to have gotten . . . : Epstein.

a letter to the . . . : Ibid.

an Op-Ed piece . . . : Samuel Epstein, "Growth Hormones Would Endanger Milk," *Los Angeles Times*, July 27, 1989.

a defense from the . . . : Gerald B. Guest, "Response by the FDA," *Los Angeles Times*, August 1, 1989.

a scientific article . . . : Samuel Epstein, "Potential Public Health Hazards of Biosynthetic Milk Hormones," *International Journal of Health Services*, 1990, page 73.

a veterinarian named . . . : Mills.

"It used to be . . .*"* : Keith Schneider, "F.D.A. Accused of Improper Ties In Review of Drug for Milk Cows," *New York Times*, January 12, 1990.

as stated in the . . . : "FDA Surveys Not Adequate to Demonstrate Safety of Milk Supply," Report to the Chairman, Human Resources and Intergovernmental Relations Subcommittee, Committee on Government Operations, House of Representatives, U.S. General Accounting Office, November 1990.

report on rBST . . . : "Recombinant Bovine Growth Hormone: FDA Approval Should Be Withheld Until the Mastitis Issue is Resolved," Report to Congressional Requesters, U.S. General Accounting Office, August 1992.

In response to . . . : Gregory N. Racz, "FDA Panel Finds Hormone Safe for Milk," *Wall Street Journal*, April 1, 1993.

Taking a new tack . . . : Richard P. Kusserow, "Need for the Food and Drug Administration to Review Possible Improper Pre-Approval Promotional Activities," Office of the Inspector General, Department of Health and Human Services, May 1991.

"There is virtually . . ." : Philip Elmer-Dewitt, "Brave New World of Milk," *Time*, February 14, 1994.

the FDA released . . . : "Interim Guidance on the Voluntary Labeling of Milk and Milk Products from Cows That Have Not Been Treated with Recombinant Bovine Somatotropin," Department of Health and Human Services, Food and Drug Administration, Docket No. 94D-0025, February 10, 1994.

In April 1994 . . . : Keith Schneider, "Maine and Vermont Restrict Dairies' Use of a Growth Hormone," *New York Times*, April 15, 1994.

then, on appeal . . . : Stacey Chase, "Court reverses BST law: Dairy-product labeling laws in doubt," *Burlington Free Press*, August 9, 1996.

took on an . . . : "Ben & Jerry's to put anti-hormone labels on products," *Augusta Chronicle*, August 15, 1997.

The label in question . . . : Margot Roosevelt Leeds, "Got Hormones?" *Time*, December 14, 2003.

"We believe Oakhurst . . ." : Matt Wickenheiser, "Oakhurst Sued by Monsanto Over Milk Advertising," *Portland Press Herald*, July 8, 2003.

To back up . . . : "Monsanto Statement Regarding Oakhurst Dairy Inc. Filing," press release, Monsanto Corporation, July 3, 2003.

the parties settled . . . : Rachel Melcer, "Monsanto Settles Milk-Labeling Lawsuit with Small Maine Dairy," *St. Louis Dispatch*, December 25, 2003.

A 2001 survey . . . : "Public Sentiment About Genetically Modified Food," prepared by the Mellman Group and Public Opinion Strategies, for the Pew Initiative on Food and Biotechnology, March 2001.

in only 22 percent . . . : "Bovine Somatotropin," Veterinary Services, Centers for Epidemiology and Animal Health, APHIS, May 2003.

In October 1999 . . . : Dirk Brinckman, "The Regulation of rBST: The European Case," *AgBioForum*, 2000.

purported offers of bribes . . . : James Baxter, "Scientists 'pressured' to approve cattle drug: Health Canada researchers accuse firm of bribery in bid to OK 'questionable' product," *Ottawa Citizen*, October 23, 1998.

"stumbled across this . . ." : Unless otherwise noted, this and all subsequent quotations from Rick North are taken from a personal interview, January 7, 2008.

a state ballot measure . . . : Patricia Callahan, "Oregon May Require Labels on Genetic Food," *Wall Street Journal*, September 30, 2002.

failure of Measure 27(: James Mayer and Michelle Cole, "Oregon Voters Make Policy Choices at Polls," *The Oregonian*, November 6, 2002.

The first big breakthrough . . . : Rick North, "Tillamook Dairy in Oregon Resists Monsanto Pressure and Bans rBGH," press release, Oregon Physicians for Social Responsibility, February 25, 2005.

"It's funny, but . . ." : Rick North, "Tillamook Dairy in Oregon Resists Monsanto Pressure and Bans rBGH," Oregon PSR Campaign for Safe Food E-mail Update, February 25, 2005.

received more than 6,500 . . . : Personal interview with Rick North, January 7, 2008.

The company would go rBST-free . . . : Craig Harris, "Starbuck switches to milk without growth hormones," *Seattle Post Intelligencer*, January 17, 2007.

"We pride ourselves . . ." : "Statement and Q&A—Starbucks Completes its Conversion—All U.S. Company-Operated Stores Use Dairy Sourced Without the Use of rBGH," press release, Starbucks Corporation, January 2008.

announced on November 5, 2007 . . . : "FIRST AND FREE: Chipotle Mexican Grill Is Nation's First Chain to Go Entirely rBGH-Free," press release, Chipotle Mexican Grill, November 5, 2007.

Bloomberg News had . . . : Jack Kaskey, "Monsanto Loss Widens; 2008 Forecast Trails Estimates (Update 6)," *Bloomberg News Service*, October 10, 2007.

filed an arbitration claim . . . : Jack Kaskey, "Monsanto Seeks $100 Million from Novartis on Hormone," *Bloomberg News Service*, August 3, 1007.

"Clean Livin' " radio ads . . . : Chipotle's radio ad can be heard online at www.chipotle.com/#flash/ads_radio. Accessed February 18, 2008.

On October 24, 2007 . . . : "Agriculture Department Notifies Companies about False or Misleading Milk and Dairy Product Labels," press release, Pennsylvania Department of Agriculture, October 24, 2007.

"Probably the most . . ." : Amy Sung, "Groups Protest Pennsylvania Milk Law," *Supermarket News*, December 10, 2007.

"Mr. Wolff's office . . ." : Andrew Martin, "Consumers Won't Know What They're Missing," *New York Times*, November 11, 2007.

Monsanto had tried the . . . : Wickenheiser.

complaining to the FDA . . . : "Federal Agencies Advised of Misleading Milk Labels and Advertising," press release, Monsanto Company, April 3, 2007. *"the company, which . . ."* : Sung.

"This is about . . ." : Rick North, "Pennsylvania Dept. of Agriculture Strikes Down rBGH-Free Labeling," E-mail update, Physicians for Social Responsibility Safe Food Campaign, October 25, 2007.

Any time you . . . : Karen Roebuck, "State's labeling changes raise ire of industry, consumers, health experts," *Pittsburgh Tribune-Review*, November 19, 2007.

Consumer groups, the . . . : Ibid.

In the end . . . : Sung; personal interview with Rick North, January 7, 2008.

announced that by . . . : David Sterrett, "Kraft Shakes Up Dairy Market; Food giant offers line of cheese free of controversial hormone," *Crain's Chicago Business News*, January 14, 2008.

"The public has . . ." : "Governor Rendell Says Consumers Can Have Greater Confidence in Milk Labels," press release, Office of Governor Edward G. Rendell, State of Pennsylvania, January 17, 2008.

"PENNSYLVANIA—YES . . ." : Rick North, "PENNSYLVANIA—YES!" E-mail update, Physicians for Social Responsibility Safe Food Campaign, January 17, 2008.

"This is a big . . ." : Daniel Malloy, "State reverses on dairy labeling, allows hormone claim," *Pittsburgh Post-Gazette*, January 18, 2008.

Chapter 9

a company called . . . : "ProdiGene, Inc.," company overview, *Business Week* online, available at investing.businessweek.com/businessweek/research/stocks/private/snapshot.asp?privcapId=33330. Accessed February 18, 2008.

it had been awarded . . . : "Terramed, Inc. receives first U.S. patent on edible vaccine technology," *Business Wire*, April 11, 1996.

Between 1997 and 2001 . . . : "Release Permits for Pharmaceuticals, Industrials, Added Proteins for Human Consumption, or for Phytoremediation Granted or Pending by APHIS as of February 22, 2008," U.S. Department of Agriculture, Animal and Plant Health Inspection Service, available online at www.aphis.usda.gov/brs/ph_permits.html. Accessed February 22, 2008.

In the fall of . . . : John Nichols, "The Three Mile Island of Biotech?" *The Nation*, December 2, 2002.

"This is your . . . " : Ibid.

In the USDA . . . : "USDA Investigates Biotech Company for Possible Permit Violations," press release, U.S. Department of Agriculture, Animal and Plant Health Inspection Service, November 13, 2002.

In its initial . . . : Ibid.

then, in March . . . : Justin Gillis, "U.S. Will Subsidize Cleanup of Altered Corn," *Washington Post*, March 26, 2003.

had been broken . . . : Scott Kilman and Sarah Lueck, "Kraft Recall Focuses on Biotechnology Oversight," *Wall Street Journal*, September 25, 2000; see also James Cox, "StarLink fiasco wreaks havoc in the heartland: Developer wants EPA to approve seed for food supply," *USA Today*, October 27, 2000.

The first incident . . . : Ibid.

Kraft spoke out . . . : Kilman and Lueck.

"Kraft's suggestions have . . ." : Ibid.

By early the . . . : Melinda Fulmer, "Kellogg says discovery of genetically modified ingredients was an isolated incident," *Los Angeles Times*, March 8, 2001.

forced to admit . . . : "Biotech Firm Executive Says Genetically Engineered Corn is Here to Stay," *Knight Ridder/Tribune*, March 19, 2001.

an optometrist in . . . : Andrew Pollack, "1999 Survey on Gene-Altered Corn Disclosed Some Improper Uses," *New York Times*, September 4, 2001.

issued a press release . . . : "USDA Investigates Biotech Company for Possible Permit Violations."

had been promising that . . . : Aaron Zitner, "Fields of Gene Factories," *Los Angeles Times*, June 4, 2001.

In his wake . . . : Mark Kawar, "Troubled Texas-Based Biotech Firm Prodi-Gene Names New CEO," *Omaha World-Herald*, March 6, 2003.

Soon thereafter . . . : Dale Johnson, "Stine Seed purchases biotech company ProdiGene," *The Spokesman*, Iowa Farm Bureau, August 25, 2003.

announced that several . . . : Justin Gillis, "Biotech Industry Adopts Precaution," *Washington Post*, October 22, 2002.

Food and grocery manufacturers . . . : Scott Kilman, "Food, Biotech Industries Feud Over Plans for Bio-Pharming," *Wall Street Journal*, November 5, 2002.

Then, in the fall . . . : Bill Freese and Richard Caplan, "Plant-Made Pharmaceuticals (PMPs) Financial Risk Profile," Report, Friends of the Earth and U.S. Public Interest Research Group, 2006.

would receive a . . . : "Turnaround of the Year Award Recipients to be Honored at New York Convention," press release, Turnaround Management Association, September 30, 2004.

its website disabled . . . : Verified at www.prodigene.com and 979–690–8537 as of February 22, 2008.

the USDA promised . . . : Griff Witte, " 'Biopharming' Bounces Back to Life," *Washington Post*, June 2. 2004.

order the USDA . . . : Sean Ho, "USDA told to disclose 'biopharm' locations," *Honolulu Advertiser*, August 5, 2004.

the USDA's own . . . : "Animal and Plant Health Inspection Service Controls Over Issuance of Genetically Engineered Organism Release Permits," Audit Report 50601–8-Te, December 2005.

a federal judge ruled . . . : Jan TenBruggencate, "Ruling a slight setback for 'biopharm' growers," *Honolulu Advertiser*, August 15, 2006.

"the agency entrusted . . ." : Ibid.

quickly and soundly rejected . . . : "California Stalls 'Pharmaceutical' Rice," *Chemical Week*, April 28, 2004.

Missouri, where . . . : Alexei Barrionuevo, "Biotech Plan in Missouri Suffers Setback," *New York Times*, January 6, 2006.

finally settled down . . . : "North Sacramento-based Ventria Sowing a storm with altered rice," *Sacramento Bee*, November 4, 2007.

who backed more than . . . : Ibid.

While promising the . . . : Ibid.

a research article . . . : Brooke D. Humphrey, Ning Huang, and Kirk C. Klasing, "Rice Expressing Lactoferrin and Lysozyme Has Antibiotic-Like Properties When Fed to Chicks," *Journal of Nutrition*, 2002.

Senate Bill 570 . . . : Text of the bill can be obtained online at pub.das .state.or.us/LEG_BILLS/PDFs_2001/ESB570.pdf. Accessed February 22, 2008.

signed into law . . . : "Summary of Legislation," Legislative Administration Committee Services, Oregon Legislative Assembly, 2007.

The committee recommended . . . : "Oregon Biopharmaceutical Committee: Policy Statement and Recommendations," October 30, 2006, available online at www.oregon.gov/ODA/PLANT/bp_policy.shtml. Accessed February 22, 2008.

In the United States . . . : "Global energy: Stirring in the corn fields," *The Economist*, May 16, 2005.

wrestled with the ethics . . . : Robin Maynard and Pat Thomas, "The next genetic revolution?" *The Ecologist*, March 29, 2007.

a letter to the . . . : C. Neal Stewart, "Biofuels and biocontainment," *Nature Biotechnology*, 2007, page 283.

the large majority . . . : Saharah Moon Chapotin and Jeffrey D. Wolt, "Genetically modified crops for the bioeconomy: meeting public and regulatory expectations," *Transgenic Research*, 2007, page 675.

would never come close . . . : Charlotte Schubert, "Can biofuels finally take center stage?" *Nature Biotechnology*, 2006, page 777.

Lignin stands second . . . : Wout Boerjan, John Ralph, and Marie Baucher, "Lignin Biosynthesis," *Annual Review of Plant Biology*, June 2003, page 519.

two of the most . . . : Chapotin and Wolt.

Switchgrass has historically . . . : "Switchgrass Burned for Power," press release, Natural Resources Conservation Service, U.S. Department of Agriculture, April 10, 2006.

long been at work . . . : Steven H. Strauss, Stephen DiFazio, and Richard Meilan, "Genetically modified poplars in context," *The Forestry Chronicle*, March/April 2001, page 271.

exceeds 80 million . . . : Vincent L. Chiang, "From rags to riches," *Nature Biotechnology*, 2002, page 557.

met with fierce . . . : Andrew Pollack, "Through Genetics, Tapping a Tree's Potential as a Source of Energy," *New York Times*, November 20, 2007; Zach Dundas, "Frankentrees: Scientists and activists face off over genetically engineered trees," *Willamette Week*, November 5, 2003.

harness the once-feared . . . : "Gut reaction fuels biofuel research," *Nature Biotechnology*, January 2008.

Conclusion

Adoption of such crops . . . : Adoption of Genetically Engineered Crops in the U.S. Data Set, U.S. Department of Agriculture Economic Research Service, July 5, 2007.

drop the release of . . . : Andrew Pollack, "Monsanto Shelves Plan For Modified Wheat," *New York Times*, May 11, 2004.

earlier been forced . . . : Scott Kilman, "Monsanto Co. Shelves Seed That Turned Out to Be a Dud of a Spud," *Wall Street Journal*, March 21, 2001.

In early 2008 . . . : Toby Van Fleet, "Modified Sweet Beet Seeds Leave Many Sour," *Portland Tribune*, March 12, 2008.

approved the planting . . . : Release Permits for Pharmaceuticals, Industrials, Value Added Proteins for Human Consumption, or for Phytoremediation Granted or Pending by APHIS as of February 26, 2008, available online and updated daily at www.aphis.usda.gov/brs/ph_permits.html. Accessed on February 26, 2008.

engineered to produce . . . : Availability of an Environmental Assessment for a Proposed Field Release of Genetically Engineered Safflower, Department of Agriculture, Animal and Plant Health Inspection Service, Docket No. APHIS-2006–0190.

field release of another . . . : SymBioSys Genetics, Inc.; Availability of an Environmental Assessment for a Field Release of Safflower Genetically Engineered To Produce Human Proinsulin, Department of Agriculture, Animal and Plant Health Inspection Service, Docket No. APHIS 2007–0023.

Germany approved . . . : "Germany gives green light to label designating 'GM free' foods," *Sydney Morning Herald*, February 16, 2008.

Index